Physics of Magnetism and Magnetic Materials

Physics of Magnetism and Magnetic Materials

K. H. J. Buschow
Van der Waals-Zeeman Instituut
Universiteit van Amsterdam
Amsterdam, The Netherlands

and

F. R. de Boer
Van der Waals-Zeeman Instituut
Universiteit van Amsterdam
Amsterdam, The Netherlands

Kluwer Academic / Plenum Publishers
New York, Boston, Dordrecht, London, Moscow

Library of Congress Cataloging-in-Publication Data

Buschow, K .H. J.
    Physics of magnetism and magnetic materials/K.H.J. Buschow & F.R. de Boer.
      p. cm.
    Includes bibliographical references and index.
    ISBN 0-306-47421-2
    1. Magnetism. 2. Magnetic materials. I. Boer, F. R. de (Frank R.) II. Title.

QC753.2 .B88 2003
538—dc21

2002034118

ISBN: 0-306-47421-2

©2003 Kluwer Academic / Plenum Publishers, New York
233 Spring Street, New York, New York 10013

http://www.wkap.nl/

10 9 8 7 6 5 4 3 2 1

A C.I.P. record for this book is available from the Library of Congress

All rights reserved

No part of this book may be reproduced, stored in a retrieval system, or transmitted in any form
or by any means, electronic, mechanical, photocopying, microfilming, recording, or otherwise,
without written permission from the Publisher, with the exception of any material supplied
specifically for the purpose of being entered and executed on a computer system, for exclusive
use by the purchaser of the work.

Printed in the United States of America

Contents

Chapter 1. Introduction 1

Chapter 2. The Origin of Atomic Moments 3
    2.1. Spin and Orbital States of Electrons 3
    2.2. The Vector Model of Atoms 5

Chapter 3. Paramagnetism of Free Ions 11
    3.1. The Brillouin Function 11
    3.2. The Curie Law 13
    References 17

Chapter 4. The Magnetically Ordered State 19
    4.1. The Heisenberg Exchange Interaction and the Weiss Field 19
    4.2. Ferromagnetism 22
    4.3. Antiferromagnetism 26
    4.4. Ferrimagnetism 34
    References 41

Chapter 5. Crystal Fields 43
    5.1. Introduction 43
    5.2. Quantum-Mechanical Treatment 44
    5.3. Experimental Determination of Crystal-Field Parameters 50
    5.4. The Point-Charge Approximation and Its Limitations 52
    5.5. Crystal-Field-Induced Anisotropy 54
    5.6. A Simplified View of 4f-Electron Anisotropy 56
    References 57

Chapter 6. Diamagnetism 59
    Reference 61

Chapter 7.	Itinerant-Electron Magnetism	63
7.1.	Introduction	63
7.2.	Susceptibility Enhancement	65
7.3.	Strong and Weak Ferromagnetism	66
7.4.	Intersublattice Coupling in Alloys of Rare Earths and 3d Metals	70
	References	73

Chapter 8.	Some Basic Concepts and Units	75
	References	83

Chapter 9.	Measurement Techniques	85
9.1.	The Susceptibility Balance	85
9.2.	The Faraday Method	86
9.3.	The Vibrating-Sample Magnetometer	87
9.4.	The SQUID Magnetometer	89
	References	89

Chapter 10.	Caloric Effects in Magnetic Materials	91
10.1.	The Specific-Heat Anomaly	91
10.2.	The Magnetocaloric Effect	93
	References	95

Chapter 11.	Magnetic Anisotropy	97
	References	102

Chapter 12.	Permanent Magnets	105
12.1.	Introduction	105
12.2.	Suitability Criteria	106
12.3.	Domains and Domain Walls	109
12.4.	Coercivity Mechanisms	112
12.5.	Magnetic Anisotropy and Exchange Coupling in Permanent-Magnet Materials Based on Rare-Earth Compounds	115
12.6.	Manufacturing Technologies of Rare-Earth-Based Magnets	119
12.7.	Hard Ferrites	122
12.8.	Alnico Magnets	124
	References	128

Chapter 13.	High-Density Recording Materials	131
13.1.	Introduction	131
13.2.	Magneto-Optical Recording Materials	133
13.3.	Materials for High-Density Magnetic Recording	139
	References	145

Chapter 14.	Soft-Magnetic Materials	147
14.1.	Introduction	147
14.2.	Survey of Materials	148
14.3.	The Random-Anisotropy Model	156
14.4.	Dependence of Soft-Magnetic Properties on Grain Size	158
14.5.	Head Materials and Their Applications	159
	14.5.1 High-Density Magnetic-Induction Heads	159
	14.5.2 Magnetoresistive Heads	161
	References	163

Chapter 15.	Invar Alloys	165
	References	170

Chapter 16.	Magnetostrictive Materials	171
	References	175

Author Index 177

Subject Index 179

1

Introduction

The first accounts of magnetism date back to the ancient Greeks who also gave magnetism its name. It derives from Magnesia, a Greek town and province in Asia Minor, the etymological origin of the word "magnet" meaning "the stone from Magnesia." This stone consisted of magnetite (Fe_3O_4) and it was known that a piece of iron would become magnetized when rubbed with it.

More serious efforts to use the power hidden in magnetic materials were made only much later. For instance, in the 18th century smaller pieces of magnetic materials were combined into a larger magnet body that was found to have quite a substantial lifting power. Progress in magnetism was made after Oersted discovered in 1820 that a magnetic field could be generated with an electric current. Sturgeon successfully used this knowledge to produce the first electromagnet in 1825. Although many famous scientists tackled the phenomenon of magnetism from the theoretical side (Gauss, Maxwell, and Faraday) it is mainly 20th century physicists who must take the credit for giving a proper description of magnetic materials and for laying the foundations of modern technology. Curie and Weiss succeeded in clarifying the phenomenon of spontaneous magnetization and its temperature dependence. The existence of magnetic domains was postulated by Weiss to explain how a material could be magnetized and nevertheless have a net magnetization of zero. The properties of the walls of such magnetic domains were studied in detail by Bloch, Landau, and Néel.

Magnetic materials can be regarded now as being indispensable in modern technology. They are components of many electromechanical and electronic devices. For instance, an average home contains more than fifty of such devices of which ten are in a standard family car. Magnetic materials are also used as components in a wide range of industrial and medical equipment. Permanent magnet materials are essential in devices for storing energy in a static magnetic field. Major applications involve the conversion of mechanical to electrical energy and vice versa, or the exertion of a force on soft ferromagnetic objects. The applications of magnetic materials in information technology are continuously growing.

In this treatment, a survey will be given of the most common modern magnetic materials and their applications. The latter comprise not only permanent magnets and invar alloys but also include vertical and longitudinal magnetic recording media, magneto optical recording media, and head materials. Many of the potential readers of this treatise may have developed considerable skill in handling the often-complex equipment of modern information technology without having any knowledge of the materials used for data storage in these systems and the physical principles behind the writing and the reading of the data. Special attention is therefore devoted to these subjects.

Although the topic Magnetic Materials is of a highly interdisciplinary nature and combines features of crystal chemistry, metallurgy, and solid state physics, the main emphasis will be placed here on those fundamental aspects of magnetism of the solid state that form the basis for the various applications mentioned and from which the most salient of their properties can be understood.

It will be clear that all these matters cannot be properly treated without a discussion of some basic features of magnetism. In the first part a brief survey will therefore be given of the origin of magnetic moments, the most common types of magnetic ordering, and molecular field theory. Attention will also be paid to crystal field theory since it is a prerequisite for a good understanding of the origin of magnetocrystalline anisotropy in modern permanent magnet materials. The various magnetic materials, their special properties, and the concomitant applications will then be treated in the second part.

2

The Origin of Atomic Moments

2.1. SPIN AND ORBITAL STATES OF ELECTRONS

In the following, it is assumed that the reader has some elementary knowledge of quantum mechanics. In this section, the vector model of magnetic atoms will be briefly reviewed which may serve as reference for the more detailed description of the magnetic behavior of localized moment systems described further on. Our main interest in the vector model of magnetic atoms entails the spin states and orbital states of free atoms, their coupling, and the ultimate total moment of the atoms.

The elementary quantum-mechanical treatment of atoms by means of the Schrödinger equation has led to information on the energy levels that can be occupied by the electrons. The states are characterized by four quantum numbers:

1. The total or principal quantum number n with values $1, 2, 3, \ldots$ determines the size of the orbit and defines its energy. This latter energy pertains to one electron traveling about the nucleus as in a hydrogen atom. In case more than one electron is present, the energy of the orbit becomes slightly modified through interactions with other electrons, as will be discussed later. Electrons in orbits with $n = 1, 2, 3, \ldots$ are referred to as occupying K, L, M, ... shells, respectively.
2. The orbital angular momentum quantum number l describes the angular momentum of the orbital motion. For a given value of l, the angular momentum of an electron due to its orbital motion equals $\hbar\sqrt{l(l+1)}$. The number l can take one of the integral values $0, 1, 2, 3, \ldots, n-1$ depending on the shape of the orbit. The electrons with $l = 0, 1, 2, 3, 4, \ldots$ are referred to as s, p, d, f, g, ... electrons, respectively. For example, the M shell ($n = 3$) can accommodate s, p, and d electrons.
3. The magnetic quantum number m_l describes the component of the orbital angular momentum l along a particular direction. In most cases, this so-called quantization direction is chosen along that of an applied field. Also, the quantum numbers m_l can take exclusively integral values. For a given value of l, one has the following possibilities: $m_l = l, l - 1, \ldots, 0, \ldots, -l + 1, -l$. For instance, for a d electron the permissible values of the angular momentum along a field direction are $2\hbar, \hbar, 0, -\hbar$, and $-2\hbar$. Therefore, on the basis of the vector model of the atom, the plane of the electronic orbit can adopt only certain possible orientations. In other words, the atom is spatially quantized. This is illustrated by means of Fig. 2.1.1.

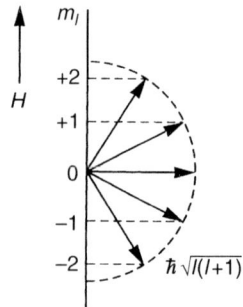

Figure 2.1.1. Vector model of the atom applied to the situation $l = 2$ and nonzero external field.

4. The spin quantum number m_s describes the component of the electron spin s along a particular direction, usually the direction of the applied field. The electron spin s is the intrinsic angular momentum corresponding with the rotation (or spinning) of each electron about an internal axis. The allowed values of m_s are $\pm 1/2$, and the corresponding components of the spin angular momentum are $\pm \hbar/2$.

According to Pauli's principle (used on p. 10) it is not possible for two electrons to occupy the same state, that is, the states of two electrons are characterized by different sets of the quantum numbers n, l, m_l, and m_s. The maximum number of electrons occupying a given shell is therefore

$$2 \sum_{l=0}^{n-1} (2l + 1) = 2n^2. \tag{2.1.1}$$

The moving electron can basically be considered as a current flowing in a wire that coincides with the electron orbit. The corresponding magnetic effects can then be derived by considering the equivalent magnetic shell. An electron with an orbital angular momentum $\hbar l$ has an associated magnetic moment

$$\vec{\mu}_l = -\frac{|e|}{2m} \hbar \vec{l} = -\mu_B \vec{l}, \tag{2.1.2}$$

where μ_B is called the Bohr magneton. The absolute value of the magnetic moment is given by

$$|\vec{\mu}_l| = \mu_B \sqrt{l(l+1)} \tag{2.1.3}$$

and its projection along the direction of the applied field is

$$\mu_{lz} = -m_l \mu_B. \tag{2.1.4}$$

The situation is different for the spin angular momentum. In this case, the associated magnetic moment is

$$\vec{\mu}_s = -g_e \frac{|e|}{2m} \hbar \vec{s} = -g_e \mu_B \vec{s}, \tag{2.1.5}$$

SECTION 2.2. THE VECTOR MODEL OF ATOMS

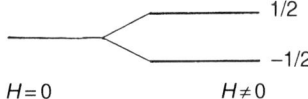

Figure 2.1.2. Effect of a magnetic field on the energy levels of the two electron states with $m_s = +1/2$ and $m_s = -1/2$.

where $g_e(=2.002290716\,(10))$ is the spectroscopic splitting factor (or the g-factor for the free electron). The component in the field direction is

$$\mu_{sz} = -g_e m_s \mu_B. \tag{2.1.6}$$

The energy of a magnetic moment $\vec{\mu}$ in a magnetic field $\vec{H}$ is given by the Hamiltonian

$$H = -\mu_0 \vec{\mu} \cdot \vec{H} = -\vec{\mu} \cdot \vec{B}, \tag{2.1.7}$$

where $\vec{B}$ is the flux density or the magnetic induction and $\mu_0 = 4\pi \times 10^{-7}\,\text{T m A}^{-1}$ is the vacuum permeability. The lowest energy E_0, the ground-state energy, is reached for $\vec{\mu}$ and $\vec{H}$ parallel. Using Eq. (2.1.6) and $m_s = -1/2$, one finds for one single electron

$$E_0 = -\mu_0 \mu_{sz} H = +g_e m_s \mu_0 \mu_B H = -\tfrac{1}{2} g_e \mu_0 \mu_B H. \tag{2.1.8}$$

For an electron with spin quantum number $m_s = +1/2$, the energy equals $+\tfrac{1}{2} g_e \mu_0 \mu_B H$. This corresponds to an antiparallel alignment of the magnetic spin moment with respect to the field.

In the absence of a magnetic field, the two states characterized by $m_s = \pm 1/2$ are degenerate, that is, they have the same energy. Application of a magnetic field lifts this degeneracy, as illustrated in Fig. 2.1.2. It is good to realize that the magnetic field need not necessarily be an external field. It can also be a field produced by the orbital motion of the electron (Ampère's law, see also the beginning of Chapter 8). The field is then proportional to the orbital angular momentum l and, using Eqs. (2.1.5) and (2.1.7), the energies are proportional to $\vec{s} \cdot \vec{l}$. In this case, the degeneracy is said to be lifted by the spin–orbit interaction.

2.2. THE VECTOR MODEL OF ATOMS

When describing the atomic origin of magnetism, one has to consider orbital and spin motions of the electrons and the interaction between them. The total orbital angular momentum of a given atom is defined as

$$\vec{L} = \sum_i \vec{l}_i, \tag{2.2.1}$$

where the summation extends over all electrons. Here, one has to bear in mind that the summation over a complete shell is zero, the only contributions coming from incomplete shells. The same arguments apply to the total spin angular momentum, defined as

$$\vec{S} = \sum_i \vec{s}_i. \qquad (2.2.2)$$

The resultants $\vec{S}$ and $\vec{L}$ thus formed are rather loosely coupled through the spin–orbit interaction to form the resultant total angular momentum $\vec{J}$:

$$\vec{J} = \vec{L} + \vec{S}. \qquad (2.2.3)$$

This type of coupling is referred to as Russell–Saunders coupling and it has been proved to be applicable to most magnetic atoms. J can assume values ranging from $J = (L-S), (L-S+1),$ to $(L+S-1), (L+S)$. Such a group of levels is called a multiplet. The level lowest in energy is called the ground-state multiplet level. The splitting into the different kinds of multiplet levels occurs because the angular momenta $\vec{L}$ and $\vec{S}$ interact with each other via the spin–orbit interaction with interaction energy $\lambda \vec{L} \cdot \vec{S}$ (λ is the spin–orbit coupling constant). Owing to this interaction, the vectors $\vec{L}$ and $\vec{S}$ exert a torque on each other which causes them to precess around the constant vector $\vec{J}$. This leads to a situation as shown in Fig. 2.2.1, where the dipole moments $\vec{\mu}_L = -\mu_B \vec{L}$ and $\vec{\mu}_S = -g_e \mu_B \vec{S}$, corresponding to the orbital and spin momentum, also precess around $\vec{J}$. It is important to realize that the total momentum $\vec{\mu}_{\text{tot}} = \vec{\mu}_L + \vec{\mu}_S$ is not collinear with $\vec{J}$ but is tilted toward the spin owing to its larger gyromagnetic ratio. It may be seen in Fig. 2.2.1 that the vector $\vec{\mu}_{\text{tot}}$ makes an angle θ with $\vec{J}$ and also precesses around $\vec{J}$. The precession frequency is usually quite high so that only the component of $\vec{\mu}_{\text{tot}}$ along $\vec{J}$ is observed, while the other component averages out to zero. The magnetic properties are therefore determined by the quantity

$$\vec{\mu} = \vec{\mu}_{\text{tot}} \cos\theta = -g_J \mu_B J. \qquad (2.2.4)$$

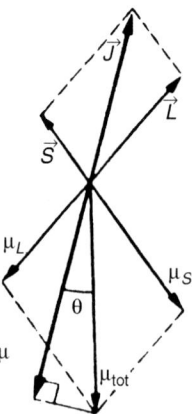

Figure 2.2.1. Spin–orbit interaction between the angular momenta $\vec{S}$ and $\vec{L}$.

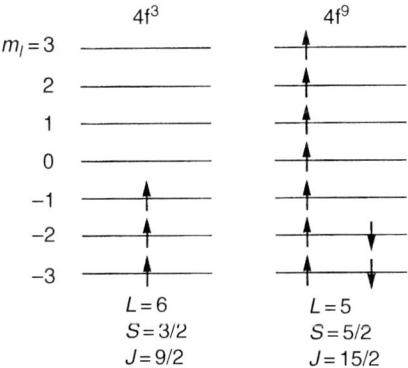

Figure 2.2.2. Application of Hund's rules to find the ground-state multiplet for an atom with three 4f electrons ($4f^3$) and nine 4f electrons ($4f^9$).

It can be shown that

$$g_J = 1 + \frac{J(J+1) + S(S+1) - L(L+1)}{2J(J+1)}. \qquad (2.2.5)$$

This factor is called the Landé spectroscopic g-factor.

For a given atom, one usually knows the number of electrons residing in an incomplete electron shell, the latter being specified by its quantum numbers. We then may use Hund's rules to predict the values of L, S, and J for the free atom in its ground state. Hund's rules are:

(1) The value of S takes its maximum as far as allowed by the exclusion principle.
(2) The value of L also takes its maximum as far as allowed by rule (1).
(3) If the shell is less than half full, the ground-state multiplet level has $J = L - S$, but if the shell is more than half full the ground-state multiplet level has $J = L + S$.

The most convenient way to apply Hund's rules is as follows. First, one constructs the level scheme associated with the quantum number l. This leads to $2l + 1$ levels, as shown for f electrons ($l = 3$) in Fig. 2.2.2. Next, these levels are filled with the electrons, keeping the spins of the electrons parallel as far as possible (rule 1) and then filling the consecutive lowest levels first (rule 2). If one considers an atom having more than $2l + 1$ electrons in shell l, the application of rule 1 implies that first all $2l + 1$ levels are filled with electrons with parallel spins before the remainder of electrons with opposite spins are accommodated in the lowest, already partly occupied, levels. Two examples of 4f-electron systems are shown in Fig. 2.2.2. The value of L is obtained from inspection of the m_l values of the occupied levels whereas S is equal to $\frac{1}{2} \times$(net number of spin-up electrons). The J values are then obtained from rule 3.

Most of the lanthanide elements have an incompletely filled 4f shell. It can be easily verified that the application of Hund's rules leads to the ground states as listed in Table 2.2.1. The variation of L and S across the lanthanide series is illustrated also in Fig. 2.2.3.

The same method can be used to find the ground-state multiplet level of the 3d ions in the iron-group salts. In this case, it is the incomplete 3d shell, which is gradually filled up.

Table 2.2.1. Selected ionic properties of the rare-earth elements. The quantity G^* represents the De Gennes factor $G = (g_J - 1)^2 J(J+1)$, normalized to the value for Gd^{3+}

Ion	$4f^n$ n	Ground term	L	S	J	g	$g\sqrt{J(J+1)}$	gJ	G^*
La^{3+}	0	$^1S_0$	0	0	—	0	0	0	0
Ce^{3+}	1	$^2F_{5/2}$	1/2	3	5/2	6/7	2.54	2.14	0.011
Pr^{3+}	2	$^3H_4$	1	5	4	4/5	3.58	3.20	0.051
Nd^{3+}	3	$^4I_{9/2}$	3/2	6	9/2	8/11	3.62	3.28	0.116
Pm^{3+}	4	$^5I_4$	2	6	4	3/5	2.68	2.40	0.217
Sm^{3+}	5	$^6H_{5/2}$	5/2	5	5/2	2/7	0.84	0.72	0.283
Eu^{3+}	6	$^7F_0$	3	3	0	0	0	0	0
Gd^{3+}	7	$^8S_{7/2}$	7/2	0	7/2	2	7.94	7	1
Tb^{3+}	8	$^7F_6$	3	3	6	3/2	9.72	9	0.667
Dy^{3+}	9	$^6H_{15/2}$	5/2	5	15/2	4/3	10.63	10	0.450
Ho^{3+}	10	$^5I_8$	2	6	8	5/4	10.60	10	0.286
Er^{3+}	11	$^4I_{15/2}$	3/2	6	15/2	6/5	9.59	9	0.162
Tm^{3+}	12	$^3H_6$	1	5	6	7/6	7.57	7	0.074
Yb^{3+}	13	$^2F_{7/2}$	1/2	3	7/2	8/7	4.54	4	0.020
Lu^{3+}	14	$^1S_0$	0	0	0	—	0	0	0

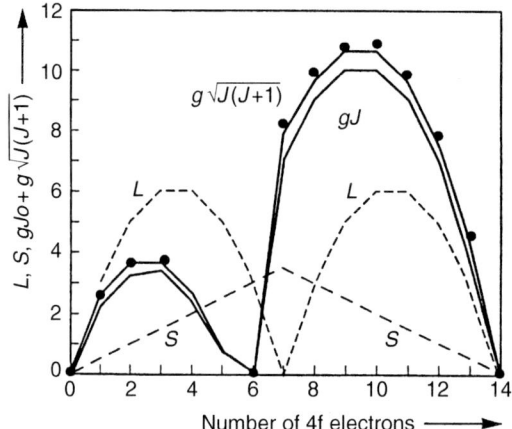

Figure 2.2.3. Variation of L, S, gJ, and $g\sqrt{J(J+1)}$ across the lanthanide series. The latter values are compared with experimental results obtained from χ^{-1} versus T plots (see Section 3.1) of various rare-earth tri-aluminides, represented as filled circles.

As seen in Tables 2.2.1 and 2.2.2, the maximum S value is reached in each case when the shells are half filled (five 3d electrons or seven 4f electrons).

In most cases, the energy separation between the ground-state multiplet level and the other levels of the same multiplet are large compared to kT. For describing the magnetic properties of the ions at 0 K, it is therefore sufficient to consider only the ground

Table 2.2.2. Selected ionic properties of iron-group elements containing Z electrons

Z	Ion	Ground term	L	S	J	$M_{\text{eff,exp}}$	$g\sqrt{J(J+1)}$	$g\sqrt{L(L+1)+4S(S+1)}$	$2\sqrt{S(S+1)}$
18	K^+, V^{5+}	$^1S_0$	0	0	0	diam.	0	0	0
19	Sc^{2+}, Ti^{3+}, V^{4+}	$^2D_{3/2}$	2	1/2	3/2	1.73	1.55	3.01	1.78
20	Ti^{2+}, V^{3+}	$^3F_2$	3	1	2	2.83	1.63	4.49	2.83
21	V^{2+}, Cr^{3+}, Mn^{4+}	$^4F_{3/2}$	3	3/2	3/2	3.82	0.70	5.21	3.87
22	Cr^{2+}, Mn^{3+}	$^5D_0$	2	2	0	4.81	0	5.50	4.91
23	Mn^{2+}, Fe^{3+}	$^6S_{5/2}$	0	5/2	5/2	5.85	5.92	5.92	5.92
24	Fe^{2+}	$^5D_4$	2	2	4	5.52 – 5.22	6.71	5.50	4.91
25	Co^{2+}	$^4F_{9/2}$	3	3/2	9/2	5.20 – 4.43	6.63	5.21	3.87
26	Ni^{2+}	$^3F_4$	3	1	4	3.23	5.59	4.49	2.83
27	Cu^{2+}	$^2D_{5/2}$	2	1/2	5/2	2.02 – 1.81	3.55	3.01	1.73

level characterized by the angular momentum quantum number J listed in Tables 2.2.1 and 2.2.2.

For completeness it is mentioned here that the components of the total angular momentum $\vec{J}$ along a particular direction are described by the magnetic quantum number m_J. In most cases, the quantization direction is chosen along the direction of the field. For practical reason, we will drop the subscript J and write simply m to indicate the magnetic quantum number associated with the total angular momentum $\vec{J}$.

3

Paramagnetism of Free Ions

3.1. THE BRILLOUIN FUNCTION

Once we have applied the vector model and Hund's rules to find the quantum numbers J, L, and S of the ground-state multiplet of a given type of atom, we can describe the magnetic properties of a system of such atoms solely on the basis of these quantum numbers and the number of atoms N contained in the system considered.

If the quantization axis is chosen in the z-direction, the z-component m of J for each atom may adopt $2J + 1$ values ranging from $m = -J$ to $m = +J$. If we apply a magnetic field H (in the positive z-direction), these $2J + 1$ levels are no longer degenerate, the corresponding energies being given by

$$E_H = -\mu_0 \vec{\mu} \cdot \vec{H} = -\mu_0 \mu_z H = g_J m \mu_0 \mu_B H, \tag{3.1.1}$$

where $\vec{\mu}$ is the atomic moment and $\mu_z = -g_J m \mu_B$ its component along the direction of the applied field $\vec{H}$ (which we have chosen as quantization direction). The constant μ_0 is equal to $4\pi \times 10^{-7}\,\text{T m A}^{-1}$.

The lifting of the $(2J+1)$-fold degeneracy of the ground-state manifold by the magnetic field is illustrated in Fig. 3.1.1 for the case $J = \frac{9}{2}$. Important features of this level scheme are that the levels are at equal distances from each other and that the overall splitting is proportional to the field strength.

Most of the magnetic properties of different types of materials depend on how this level scheme is occupied under various experimental circumstances. At zero temperature, the situation is comparatively simple because for any of the N participating atoms only the lowest level will be occupied. In this case, one obtains for the magnetization of the system

$$M = -N g_J m \mu_B = N g_J J \mu_B. \tag{3.1.2}$$

However, at finite temperatures, higher lying levels will become occupied. The extent to which this happens depends on the temperature but also on the energy separation between the ground-state level and the excited levels, that is, on the field strength.

The relative population of the levels at a given temperature T and a given field strength H can be determined by assuming a Boltzmann distribution for which the probability P_i of

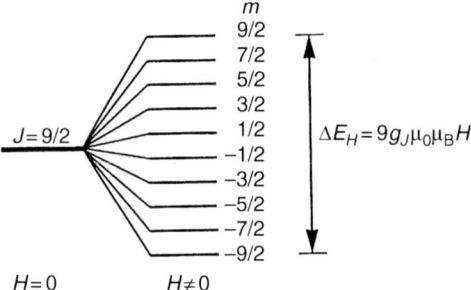

Figure 3.1.1. Splitting of the $(2J+1)$-fold degenerate $J = 9/2$ level under the action of a magnetic field. The overall splitting ΔE_H is proportional to the field strength H.

finding an atom in a state with energy E_i is given by

$$P_i = \frac{\exp(-E_i/kT)}{\sum_i \exp(-E_i/kT)}. \tag{3.1.3}$$

The magnetization M of the system can then be found from the statistical average $\langle \mu_z \rangle$ of the magnetic moment $\mu_z = -g_J m \mu_B$. This statistical average is obtained by weighing the magnetic moment μ_z of each state by the probability that this state is occupied and summing over all states:

$$M = N\langle\mu_z\rangle = N \frac{\sum_{m=-J}^{J} -g_J m \mu_B \exp(-g_J m \mu_0 \mu_B H/kT)}{\sum_{m=-J}^{J} \exp(-g_J m \mu_0 \mu_B H/kT)}. \tag{3.1.4}$$

The calculation of the magnetization by means of this formula is a cumbersome procedure and eventually leads to Eq. (3.1.10). For the readers who are interested in how this result has been reached and in the approximations made, a simple derivation is given below. Since there is no magnetism but merely algebra involved in this derivation, the average reader will not lose much when jumping directly to Eq. (3.1.10), keeping in mind that the magnetization given by Eq. (3.1.10) is a result of the thermal averaging in Eq. (3.1.4), involving $2J + 1$ equidistant energy levels.

By substituting $x = -g_J \mu_B \mu_0 H/kT$ into Eq. (3.1.4), and using the relations $d \ln x = x^{-1} dx$ and $de^{mx} = me^{mx} dx$, one may write

$$M = Ng\mu_B \frac{d}{dx}\left(\ln \sum_{m=-J}^{J} e^{mx}\right). \tag{3.1.5}$$

Since there cannot be any confusion with g_e here, we have dropped the subscript J of g_J and simply write g from now on.

From the standard expression for the sum of a geometric series, one finds

$$\sum_{m=-J}^{J} e^{mx} = e^{-Jx}(1 + e^x + e^{2x} + \cdots + e^{2Jx}) = e^{-Jx}\frac{e^{(2J+1)x} - 1}{e^x - 1}. \tag{3.1.6}$$

Substitution of this result into Eq. (3.1.5) leads to

$$M = Ng\mu_B \frac{d}{dx}\left(\ln \frac{e^{(J+1)x} - e^{-Jx}}{e^x - 1}\right)$$

$$= Ng\mu_B \frac{d}{dx}\left(\ln \frac{e^{\left(J+\frac{1}{2}\right)x} - e^{-\left(J+\frac{1}{2}\right)x}}{e^{\frac{1}{2}x} - e^{-\frac{1}{2}x}}\right). \tag{3.1.7}$$

Since $\sinh x = (e^x - e^{-x})/2$, one obtains

$$M = Ng\mu_B \frac{d}{dx}\left(\ln \frac{\sinh\left(J+\frac{1}{2}\right)x}{\sinh \frac{1}{2}x}\right). \tag{3.1.8}$$

After carrying out the differentiation, one finds

$$M = Ng\mu_B J B_J(y), \tag{3.1.9}$$

with $B_J(y)$, the so-called Brillouin function, given by

$$B_J(y) = \frac{2J+1}{2J}\coth\frac{(2J+1)y}{2J} - \frac{1}{2J}\coth\frac{y}{2J}, \tag{3.1.10}$$

with

$$y = \frac{gJ\mu_B\mu_0 H}{kT}. \tag{3.1.11}$$

It is good to bear in mind that in this expression H is the field responsible for the level splitting of the $2J + 1$ ground-state manifold. In most cases, H is the externally applied magnetic field. We shall see, however, in one of the following chapters that in some materials also internal fields are present which may cause the level splitting of the $(2J+1)$-manifold.

Expression (3.1.9) makes it possible to calculate the magnetization for a system of N atoms with quantum number J at various combinations of applied field and temperature.

Experimental results for the magnetization of several paramagnetic complex salts containing Cr^{3+}, Fe^{3+}, and Gd^{3+} ions measured in various field strengths at low temperatures are shown in Fig. 3.1.2. The curves through the data points have been calculated by means of Eq. (3.1.9). There is good agreement between the calculations and the experimental data.

3.2. THE CURIE LAW

Expression (3.1.9) becomes much simpler in cases where the temperature is higher and the field strength lower than for most of the data shown in Fig. 3.1.2. In order to see this, we will assume that we wish to study the magnetization at room temperature of a complex salt of Nd^{3+} in an external field $H = 80$ kA m^{-1}, which corresponds to an external flux density $B = \mu_0 H = 0.1$ T ($\mu_0 = 4\pi \times 10^{-7}$ T/A m^{-1}; more details about units will be discussed

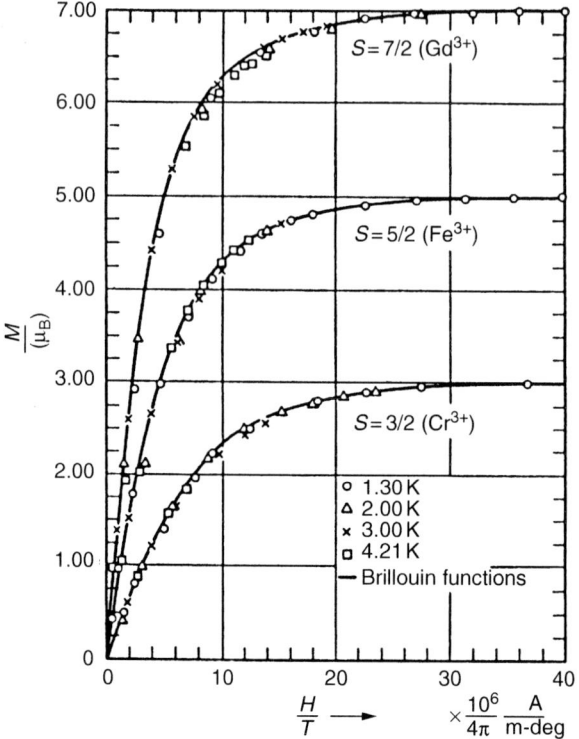

Figure 3.1.2. Magnetization M (in μ_B/atom) of several paramagnetic complex salts containing Gd^{3+}, Fe^{3+}, and Cr^{3+} plotted versus $\mu_0 H/T$ (in T/K). After Henry (1952).

in Chapter 8). For Nd^{3+}, one has $J = 9/2$ and $g = 8/11$ (see Table 2.2.1). Furthermore, we make use of the following values

$$\mu_B = 9.274 \times 10^{-24} \text{ A m}^2 \text{ (or J/T)} \quad \text{and} \quad k = 1.381 \times 10^{-23} \text{ J/K}.$$

At room temperature (298 K), one derives for y in Eq. (3.1.11):

$$y = \frac{(9/2) \cdot (8/11) \cdot 9.27 \cdot 10^{-24} \cdot 0.1}{1.38 \cdot 10^{-23} \cdot 298} = 0.00074.$$

Since we now have shown that $y \ll 1$ under the above conditions, it is justified to use only the first term of the series expansion of $B_J(y)$ for small values of y

$$B_J(y) = \frac{J+1}{3J} y - \frac{[(J+1)^2 + J^2](J+1)}{90 J^3} y^3 + \cdots \quad (3.2.1)$$

From this follows, keeping only the first term,

$$M = N g \mu_B J B_J(y) = N g \mu_B J \cdot \frac{J+1}{3J} \cdot \frac{J g \mu_B \mu_0 H}{kT} = \frac{N \mu_0 g^2 J(J+1) \mu_B^2 H}{3kT}. \quad (3.2.2)$$

SECTION 3.2. THE CURIE LAW

The magnetic susceptibility is defined as $\chi = M/H$. Using Eq. (3.2.2), we derive for the magnetic susceptibility

$$\chi = \frac{N\mu_0 g^2 J(J+1)\mu_B^2}{3kT} = \frac{C}{T}, \quad (3.2.3)$$

with the Curie constant C given by

$$C = \frac{N\mu_0 g^2 J(J+1)\mu_B^2}{3k}. \quad (3.2.4)$$

Relationship (3.2.3) is known as the Curie's law because it was first discovered experimentally by Curie in 1895. Curie's law states that if the reciprocal values of the magnetic susceptibility, measured at various temperatures, are plotted versus the corresponding temperatures, one finds a straight line passing through the origin. From the slope of this line one finds a value for the Curie constant C and hence a value for the effective moment

$$\mu_{\text{eff}} = g\sqrt{J(J+1)}\mu_B. \quad (3.2.5)$$

The Curie behavior may be illustrated by means of results of measurements made on the intermetallic compound TmAl$_3$ shown in Fig. 3.2.1.

It is seen that the reciprocal susceptibility is linear over almost the whole temperature range. From the slope of this line one derives $\mu_{\text{eff}} = 7.68$ μ_B per Tm atom, which is close to the value expected on the basis of Eq. (3.2.5) with J and g determined by Hund's rules (values listed in Table 2.2.1). Similar experiments made on most of the other types of rare-earth tri-aluminides also lead to effective moments that agree closely with the values derived with Eq. (3.2.5). This may be seen from Fig. 2.2.3 where the upper full line represents the variation of $g\sqrt{J(J+1)}$ across the rare-earth series and where the effective moments experimentally observed for the tri-aluminides are given as full circles. In all these cases, one has a situation basically the same as that shown in the inset of Fig. 3.2.1 for Tm^{3+}, where the ground-state multiplet level lies much lower than the first excited multiplet level. In these cases, one needs to take into account only the $2J+1$ levels of the ground-state multiplet, as we did when calculating the statistical average by means of Eq. (3.1.4). Note that in the temperature range considered in Fig. 3.2.1, the first excited level $J=4$ will practically not be populated.

The situation is different, however, for Sm^{3+} and Eu^{3+}. It is shown in the inset of Fig. 3.2.1 that for Sm^{3+} several excited multiplet levels occur which are not far from the ground state. Each of these levels will be split by the applied magnetic field into $2J+1$ sublevels. At very low temperatures, only the $2J+1$ levels of the ground-state multiplet are populated. With increasing temperature, however, the sublevels of the excited states also become populated. Since these levels have not been considered in the derivation of Eq. (3.2.3) via Eq. (3.1.4), one may expect that Eq. (3.2.3) does not provide the right answer here. With increasing temperature, there would have been an increasing contribution of the sublevels of the excited states to the statistical average if we had included these levels in the summation in Eq. (3.1.4). Since, for Sm^{3+}, the excited multiplet levels have higher magnetic moments than the ground state, one expects that M and χ will increase with increasing temperature for sufficiently high temperatures. This means that χ^{-1} will decrease with increasing temperature, which is a strong violation of the Curie law (Eq. 3.2.3). Experimental results for SmAl$_3$ demonstrating this exceptional behavior are shown in Fig. 3.2.1.

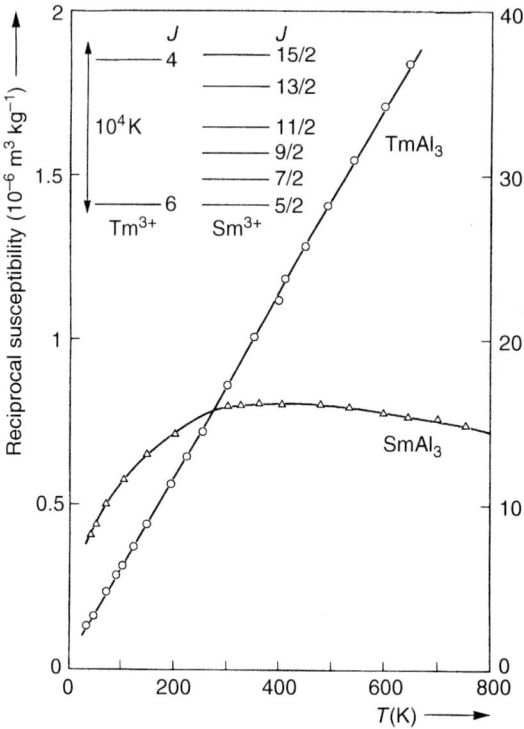

Figure 3.2.1. Temperature dependence of the reciprocal susceptibility measured on TmAl$_3$ (left scale) and SmAl$_3$ (right scale). The inset shows that the ground-state multiplet ($J = 6$) is well isolated for Tm^{3+}, the first excited multiplet level ($J = 4$) lying about 10,000 K higher. For Sm^{3+}, the ground-state multiplet $J = 5/2$ is less well-separated from the excited multiplets.

The magnetic splitting of the ground-state multiplet level ($J = L - S = 5 - 5/2 = 5/2$) and the first excited multiplet level ($J = L - S + 1 = 5 - 5/2 + 1 = 7/2$) is illustrated in Fig. 3.2.2. Note that the equidistant character is lost not only due to the energy gap between the $J = 5/2$ and $J = 7/2$ levels but also due to a difference in energy separation between the levels of the $J = 5/2$ manifold ($g = 2/7$) and the levels of the $J = 7/2$ manifold ($g = 52/63$).

Generally speaking, it may be stated that the Curie law $\chi = C/T$, as expressed in Eq. (3.2.3), is a consequence of the fact that the thermal average calculated in Eq. (3.1.4) involves only the $2J + 1$ equally spaced levels (see Fig. 3.1.1) originating from the effect of the applied field on one multiplet level. Deviations from Curie behavior may be expected whenever more than these $2J + 1$ levels are involved (as for Sm^{3+} and Eu^{3+}), or when these levels are no longer equally spaced. The latter situation occurs when electrostatic fields in the solid, the crystal fields, come into play. It will be shown later how crystal fields can also lift the degeneracy of the $2J + 1$ ground-state manifold. The combined action of crystal fields and magnetic fields generally leads to a splitting of this manifold in which the $2J + 1$

SECTION 3.2. THE CURIE LAW

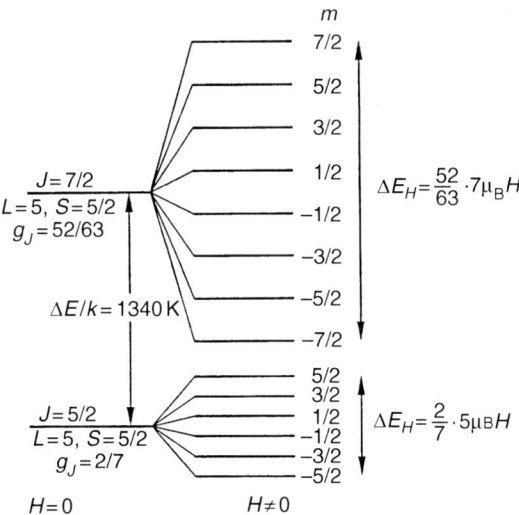

Figure 3.2.2. Schematic representation of the magnetic splitting of the ground-state multiplet ($J = 5/2$) and the first excited multiplet ($J = 7/2$) of Sm^{3+}.

sublevels are no longer equally spaced, or to a splitting where the level with $m = -J$ is not the lowest level in moderate magnetic fields.

More detailed treatments of the topics dealt with in this chapter can be found in the textbooks of Morrish (1965) and Martin (1967).

References

Henry, W. E. (1952) *Phys. Rev.*, **88**, 559.
Martin, D. H. (1967) *Magnetism in Solids*, London: Iliffe Books Ltd.
Morrish, A. H. (1965) *The Physical Principles of Magnetism*, New York: John Wiley and Sons.

4

The Magnetically Ordered State

4.1. THE HEISENBERG EXCHANGE INTERACTION AND THE WEISS FIELD

It follows from the results described in the previous sections, that all N atomic moments of a system will become aligned parallel if the conditions of temperature and applied field are such that for all of the participating magnetic atoms only the lowest level ($m = -J$ in Fig. 3.1.1) is occupied. The magnetization of the system is then said to be saturated, no higher value being possible than

$$M_s = Ng\mu_B J. \quad (4.1.1)$$

This value corresponds to the horizontal part of the three magnetization curves shown in Fig. 3.1.2. It may furthermore be seen from Fig. 3.1.2 that the parallel alignment of the moments is reached only in very high applied fields and at fairly low temperatures. This behavior of the three types of salts represented in Fig. 3.1.2 strongly contrasts the behavior observed in several normal magnetic metals such as Fe, Co, Ni, and Gd, in which a high magnetization is already observed even without the application of a magnetic field. These materials are called ferromagnetic materials and are characterized by a spontaneous magnetization. This spontaneous magnetization vanishes at temperatures higher than the so-called Curie temperature T_C. Below T_C, the material is said to be ferromagnetically ordered.

On the basis of our understanding of the magnetization in terms of the level splitting and level population discussed in the previous section (Eq. 3.1.4; Fig. 3.1.1), the occurrence of spontaneous magnetization would be compatible with the presence of a huge internal magnetic field, H_m. This internal field should then be able to produce a level splitting of sufficient magnitude so that practically only the lowest level $m = -J$ is populated. Heisenberg has shown in 1928 that such an internal field may arise as the result of a quantum-mechanical exchange interaction between the atomic spins. The Heisenberg exchange Hamiltonian is usually written in the form

$$H_{\text{exch}} = \sum_{i<j} 2J_{ij} \vec{S}_i \cdot \vec{S}_j, \quad (4.1.2)$$

where the summation extends over all spin pairs in the crystal lattice. The exchange constant J_{ij} depends, amongst other things, on the distance between the two atoms i and j considered.

In most cases, it is sufficient to consider only the exchange interaction between spins on nearest-neighbor atoms. If there are Z magnetic nearest-neighbor atoms surrounding a given magnetic atom, one has

$$H_{\text{exch}} = -2ZJ_{\text{nn}}\vec{S} \cdot \langle \vec{S} \rangle, \tag{4.1.3}$$

with $\langle \vec{S} \rangle$ the average spin of the nearest-neighbor atoms. Relation (4.1.3) can be rewritten by using $\vec{S} = (g-1)\vec{J}$, which follows from the relations $g\vec{J} = \vec{L} + 2\vec{S}$ and $\vec{J} = \vec{L} + \vec{S}$ (Fig. 2.1.2):

$$H_{\text{exch}} = -2ZJ_{\text{nn}}(g-1)^2 \vec{J} \cdot \langle \vec{J} \rangle. \tag{4.1.4}$$

Since the atomic moment is related to the angular momentum by $\vec{\mu} = -g_J \mu_B \vec{J}$ (Eq. 2.2.4), we may also write

$$H_{\text{exch}} = \frac{-2ZJ_{\text{nn}}(g-1)^2 \vec{\mu} \cdot \langle \vec{\mu} \rangle}{g^2 \mu_B^2} = -\mu_0 \vec{\mu} \cdot \vec{H}_m, \tag{4.1.5}$$

where

$$\vec{H}_m = \frac{2ZJ_{\text{nn}}(g-1)^2 \langle \vec{\mu} \rangle}{g^2 \mu_B^2} \tag{4.1.6}$$

can be regarded as an effective field, the so-called molecular field, produced by the average moment $\langle \vec{\mu} \rangle$ of the Z nearest-neighbor atoms.

Since $M = N \langle \vec{\mu} \rangle$, it follows furthermore that $\vec{H}_m$ is proportional to the magnetization

$$\vec{H}_m = N_W \vec{M}. \tag{4.1.7}$$

The constant N_W is called the molecular-field constant or the Weiss-field constant. In fact, Pierre Weiss postulated the presence of a molecular field in his phenomenological theory of ferromagnetism already in 1907, long before its quantum-mechanical origin was known.

The exchange interaction between two neighboring spin moments introduced in Eq. (4.1.2) has the same origin as the exchange interaction between two electrons on the same atom, where it can lead to parallel and antiparallel spin states. The exchange interaction between two neighboring spin moments arises as a consequence of the overlap between the magnetic orbitals of two adjacent atoms. This so-called direct exchange interaction is strong in particular for 3d metals, because of the comparatively large extent of the 3d-electron charge cloud. Already in 1930, Slater found that a correlation exists between the nature of the exchange interaction (sign of exchange constant in Eq. 4.1.2) and the ratio r_{ab}/r_d, where r_{ab} represents the interatomic distance and r_d the radius of the incompletely filled d shell. Large values of this ratio corresponded to a positive exchange constant, while for small values it was negative.

Quantum-mechanical calculations based on the Heitler–London approach were made by Sommerfeld and Bethe (1933). These calculations largely confirmed the result of Slater and have led to the Bethe–Slater curve shown in Fig. 4.1.1. According to this curve, the exchange interaction between the moments of two similar 3d atoms changes when these are brought closer together. It is comparatively small for large interatomic distances, passes through a maximum, and eventually becomes negative for rather small interatomic distances. As indicated in the figure, this curve has been most successful in separating the

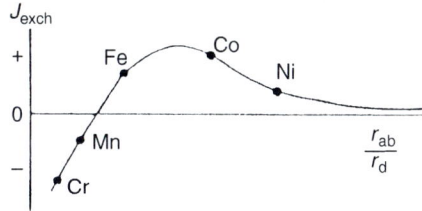

Figure 4.1.1. Bethe–Slater curve describing the variation of the exchange constant with interatomic separation r_{ab} and radius r_d of the incompletely filled shell.

Table 4.1.1. Comparison of shortest Fe–Fe distances d and magnetic-ordering temperatures found in α-Fe and various types of binary and ternary Fe-based intermetallic compounds. The corresponding Fe sites in the various structure types have been indicated in Wyckof notation

Compound	Structure type	Fe sites	Distance d (Å)	Ordering temperature (K)
$NdFe_{12}B_6$	$SrNi_{12}B_6$	18h–18h	2.38	$T_C = 230$
Ce_2Fe_{17}	Ce_2Fe_{17}	6c–6c	2.38	$T_N = 225$
Lu_2Fe_{17}	Th_2Ni_{17}	4f–4f	2.39	$T_N = 270$
$Nd_2Fe_{14}B$	$Nd_2Fe_{14}B$	$16k_2$–$8j_1$	2.40	$T_C = 588$
$Nd_2Fe_{23}B_3$	$Nd_2Fe_{23}B_3$	48e–24d	2.40	$T_C = 655$
$YFe_{11}Ti$	$ThMn_{12}$	8f–8f	2.40	$T_C = 524$
Fe_2B	Al_2Cu	8h–8h	2.45	$T_C = 1013$
α-Fe	W	2a–2a	2.48	$T_C = 1044$
FeGe	B 35	3f–3f	2.50	$T_N = 412$
$HfFe_6Ge_6$	$HfFe_6Ge_6$	6i–6l	2.51	$T_N = 453$
YFe_4Al_8	$ThMn_{12}$	8f–8f	2.52	$T_N = 185$
$LuFe_6Ge_6$	$HfFe_6Ge_6$	6i–6l	2.55	$T_N = 452$
$FePt_3$	Cu_3Au	1a–1a	3.78	$T_N = 170$

ferromagnetic 3d elements like Ni, Co, and Fe (parallel moment arrangements) from the antiferromagnetic elements Mn and Cr (antiparallel moment arrangements).

The validity of the Bethe–Slater curve has seriously been criticized by several authors. As discussed by Herring (1966), this curve lacks a sound theoretical basis. In the form of a semi-empirical curve, it is still widely used to explain changes in the magnetic moment coupling when the interatomic distance between the corresponding atoms is increased or decreased. Even though this curve may be helpful in some cases to explain and predict trends, it should be borne in mind that it might not be generally applicable.

We will investigate this point further by looking at some data collected in Table 4.1.1. In this table, magnetic-ordering temperatures are listed for ferromagnetic compounds (T_C) and antiferromagnetic compounds (T_N). As will be explained in the following sections, negative exchange interactions leading to antiparallel moment coupling exist in the latter compounds. The shortest interatomic Fe–Fe distances occurring in the corresponding crystal structures have also been included in Table 4.1.1. The shortest Fe–Fe distances, for which antiferromagnetic couplings are predicted to occur according to Fig. 4.1.1, are seen to adopt a wide gamut of values on either side of the Fe–Fe distance in Fe metal.

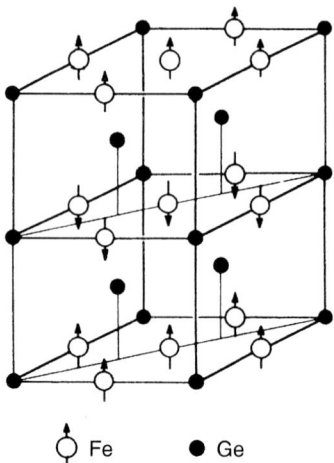

Figure 4.1.2. Arrangement of Fe moments (arrows) in the hexagonal compound FeGe. The Ge atoms are indicated by the filled circles. After Beckman and Lundgren (1991).

This does not lend credence to the notion that short Fe–Fe distances favor antiferromagnetic interactions. Equally illustrative in this respect is the magnetic moment arrangement in the compound FeGe shown in Fig. 4.1.2. The shortest Fe–Fe distance (2.50 Å) occurring in the horizontal planes gives rise to ferromagnetic rather than antiferromagnetic interaction. Antiferromagnetic interaction occurs between Fe moments separated by much larger distances (4.05 Å) along the vertical direction. This is a behavior opposite to that expected on the basis of the Bethe–Slater curve, showing that its validity is rather limited.

4.2. FERROMAGNETISM

The total field experienced by the magnetic moments comprises the applied field H and the molecular field or Weiss field H_m:

$$\vec{H}_{tot} = \vec{H} + \vec{H}_m = \vec{H} + N_W \vec{M}. \tag{4.2.1}$$

We will first investigate the effect of the presence of the Weiss field $N_W M$ on the magnetic behavior of a ferromagnetic material above T_C. In this case, the magnetic moments are no longer ferromagnetically ordered and the system is paramagnetic. Therefore, we may use again the high-temperature approximation by means of which we have derived Eq. (3.2.2)

$$M = \frac{C}{T} H. \tag{4.2.2}$$

We have to bear in mind, however, that the splitting of the $(2J + 1)$-manifold used to calculate the statistical average $\langle \mu_z \rangle$ is larger owing to the presence of the Weiss field. For a ferromagnet above T_C we therefore have to use H_{tot} instead of H when going through

SECTION 4.2. FERROMAGNETISM

all the steps from Eq. (3.1.4) to Eq. (3.2.2). This means that Eq. (3.2.2) should actually be written in the form

$$M = \frac{C}{T}(H + N_W M). \qquad (4.2.3)$$

Introducing the magnetic susceptibility $\chi = M/H$, we may rewrite Eq. (4.2.3) into

$$\chi = \frac{C}{T - N_W C} = \frac{C}{T - \theta_p}, \qquad (4.2.4)$$

where θ_p is called the asymptotic or paramagnetic Curie temperature.

Relation (4.2.4) is known as the Curie–Weiss law. It describes the temperature dependence of the magnetic susceptibility for temperatures above T_C. The reciprocal susceptibility when plotted versus T is again a straight line. However, this time it does not pass through the origin (as for the Curie law) but intersects the temperature axis at $T = \theta_p$. Plots of χ^{-1} versus T for an ideal paramagnet ($\chi = C/T$) and a ferromagnetic material above T_C ($\chi = C/(T - \theta_p)$) are compared with each other in Fig. 4.2.1.

One notices that at $T = \theta_p$, the susceptibility diverges which implies that one may have a nonzero magnetization in a zero applied field. This exactly corresponds to the definition of the Curie temperature, being the upper limit for having a spontaneous magnetization. We can, therefore, write for a ferromagnet

$$\theta_p = T_C = N_W C = \frac{N_W N \mu_0 g^2 J(J+1)\mu_B^2}{3k}. \qquad (4.2.5)$$

This relation offers the possibility to determine the magnitude of the Weiss constant N_W from the experimental value of T_C or θ_p, obtained by plotting the spontaneous magnetization versus T or by plotting the reciprocal susceptibility versus T, respectively (see Fig. 4.2.1c).

We now come to the important question of how to describe the magnetization of a ferromagnetic material below its Curie temperature. Of course, when the temperature approaches zero kelvin only the lowest level of the $(2J + 1)$-manifold will be populated and we have

$$M(T = 0) = M_s = N g \mu_B J. \qquad (4.2.6)$$

In order to find the magnetization between $T = 0$ and $T = T_C$, we have to return to Eq. (3.1.9) which we will write now in the form

$$M(T) = N g \mu_B J B_J(y) = M(0) B_J(y), \qquad (4.2.7)$$

with

$$y = \frac{g J \mu_B \mu_0 H_{tot}}{kT}, \qquad (4.2.8)$$

where H_{tot} is the total field responsible for the level splitting of the $2J + 1$ ground-state manifold.

The total magnetic field experienced by the atomic moments in a ferromagnet is $H_{tot} = H + H_m$ and, since we are interested in the spontaneous magnetization (at $H = 0$), we have to use $H_{tot} = H_m = N_W M$ (Eq. 4.1.7), or rather $H_{tot}(T) = N_W M(T)$. This means that y in Eq. (4.2.8) is now given by

$$y = \frac{g J \mu_B \mu_0 H_m}{kT} = \frac{g J \mu_B \mu_0 N_W M(T)}{kT}. \qquad (4.2.9)$$

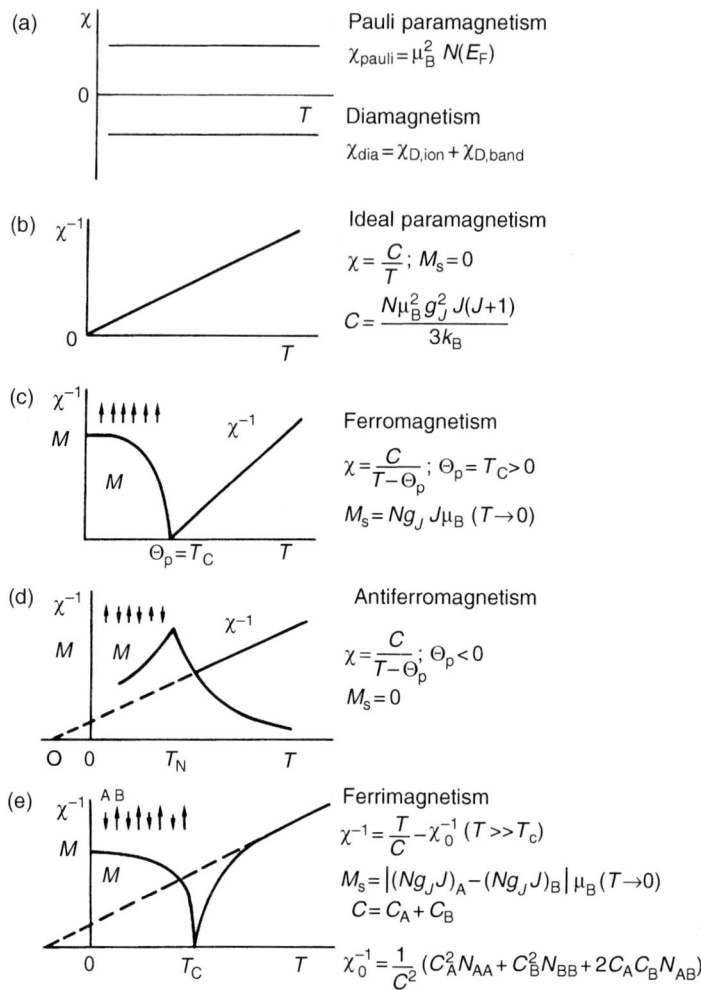

Figure 4.2.1. Summary of the temperature dependence of the magnetization M, the magnetic susceptibility χ or the reciprocal susceptibility χ^{-1} in various types of magnetic materials. After Buschow (1994).

Combining this expression with Eq. (4.2.7) leads to

$$M(T) = Ng\mu_B J B_J\left(\frac{gJ\mu_B\mu_0 N_W M(T)}{kT}\right). \quad (4.2.10)$$

Upon substitution of $N_W = T_C/C$ (Eq. 4.2.5) and $M(0) = Ng\mu_B J$ into Eq. (4.2.10), one finds

$$\frac{M(T)}{M(0)} = B_J\left(\frac{3J}{J+1}\cdot\frac{T_C}{T}\cdot\frac{M(T)}{M(0)}\right). \quad (4.2.11)$$

This is quite an interesting result because it shows that for a given J the variation of the reduced magnetization $M(T)/M(0)$ with the reduced temperature T/T_C depends

SECTION 4.2. FERROMAGNETISM

exclusively on the form of the Brillouin function B_J. It is independent of parameters that vary from one material to the other such as the atomic moment gJ, the number of participating magnetic atoms N and the actual value of T_C. In fact, the variation of the reduced magnetization with the reduced temperature can be regarded as a law of corresponding states that should be obeyed by all ferromagnetic materials. This was a major achievement of the Weiss theory of ferromagnetism, albeit Weiss, instead of using the Brillouin function, obtained this important result by using the classical Langevin function for calculating $M(T)$:

$$M(T) = M(0)L(x), \qquad (4.2.12)$$

with

$$L(x) = \coth x - \frac{1}{x} \quad \text{and} \quad x = \frac{m_0 \mu_0 H}{kT}. \qquad (4.2.13)$$

Here m_0 represents the classical atomic moment that, in the classical description, is allowed to adopt any direction with respect to the field H (no directional quantization). The classical Langevin function is obtained by calculating the statistical average $\langle m_0 \cos\theta \rangle$ of the moment m_0 in the direction of the field. A derivation of the Langevin function will not be given here. For more details, the reader is referred to the textbooks of Morrish (1965), Chikazumi and Charap (1966), Martin (1967), White (1970), and Barbara et al. (1988).

Several curves of the reduced magnetization versus the reduced temperature, calculated for the ferromagnetic Brillouin functions (Eq. 4.2.11) with $J = \frac{1}{2}$, 1, and ∞ are shown in Fig. 4.2.2, where they can be compared with experimental results of two materials with strongly different Curie temperatures: iron ($T_C = 1044$ K) and nickel ($T_C = 627$ K).

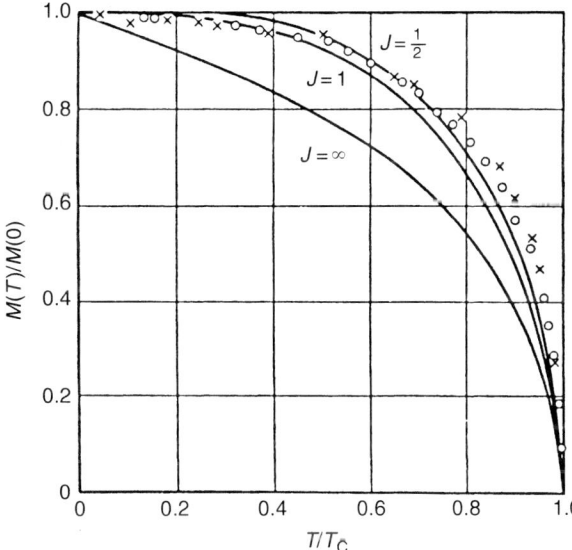

Figure 4.2.2. Reduced magnetization $M(T)/M(0)$ versus reduced temperature T/T_C calculated by means of the ferromagnetic Brillouin function for several values of J. Included are experimental data for iron (x) and nickel (o). After Becker and Döring (1939).

4.3. ANTIFERROMAGNETISM

A simple antiferromagnet can be visualized as consisting of two magnetic sublattices (A and B). In the magnetically ordered state, the atomic moments are parallel or ferromagnetically coupled within each of the two sublattices. Any two atomic magnetic moments belonging to different sublattices have an antiparallel orientation. Since the moments of both sublattices have the same magnitude and since they are oriented in opposite directions, one finds that the total magnetization of an antiferromagnet is essentially zero (at least at zero kelvin). As an example, the unit cell of a simple antiferromagnet is shown in Fig. 4.3.1.

In order to describe the magnetic properties of antiferromagnets, we may use the same concepts as in the previous section. However, it will be clear that the molecular field caused by the moments of the same sublattice will be different from that caused by the moments of the other (antiparallel) sublattice. The total field experienced by the moments of sublattices A and B can then be written as

$$\vec{H}_A = \vec{H} + N_{AA}\vec{M}_A + N_{AB}\vec{M}_B, \tag{4.3.1}$$

$$\vec{H}_B = \vec{H} + N_{BA}\vec{M}_A + N_{BB}\vec{M}_B, \tag{4.3.2}$$

where H is the external field and where the sublattice moments M_A and M_B have the same absolute value:

$$|\vec{M}_A| = |\vec{M}_B| = \tfrac{1}{2}NgJ\mu_B. \tag{4.3.3}$$

The intrasublattice-molecular-field constant $N_{AA} = N_{BB} = N_1$ is different in magnitude and sign from the intersublattice-molecular-field constant $N_{AB} = N_{BA} = N_2$.

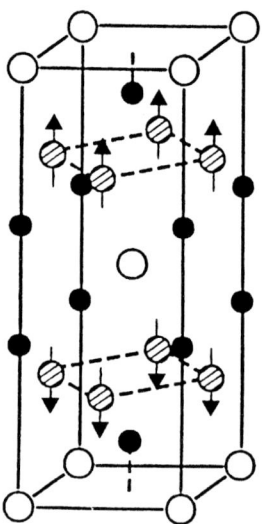

Figure 4.3.1. Arrangement of the magnetic moments in the unit cell of the antiferromagnet YMn$_2$Ge$_2$ below its Néel temperature $T_N = 395$ K. The open circles represent the Y atoms, the dashed circles the (magnetic) Mn atoms, and the black circles the Ge atoms.

SECTION 4.3. ANTIFERROMAGNETISM

The temperature dependence of each of the two sublattice moments can be obtained by means of Eq. (3.1.9):

$$M_A = M(0) B_J(y), \qquad (4.3.4)$$

with

$$y = \frac{\frac{1}{2} g J \mu_B \mu_0 |\vec{H}_A|}{kT}. \qquad (4.3.5)$$

A similar expression holds for M_B.

In analogy with Eq. (4.2.3), it is relatively easy to derive expressions for the sublattice moments in the high-temperature limit:

$$M_A = \frac{C'}{T}(H + N_1 M_A + N_2 M_B), \qquad (4.3.6)$$

$$M_B = \frac{C'}{T}(H + N_2 M_A + N_1 M_B), \qquad (4.3.7)$$

where

$$C' = C_A = C_B = \frac{1}{2} C = \frac{\frac{1}{2} N \mu_0 g^2 J(J+1) \mu_B^2}{3k}. \qquad (4.3.8)$$

The two coupled equations for M_A and M_B will lead to spontaneous sublattice moments ($|M_A| = |M_B| \neq 0$ for $H = 0$) if the determinant of the coefficients of M_A and M_B vanishes:

$$\begin{vmatrix} \frac{C' N_1}{T} - 1 & \frac{C' N_2}{T} \\ \frac{C' N_2}{T} & \frac{C' N_1}{T} - 1 \end{vmatrix} = 0. \qquad (4.3.9)$$

The temperature at which the spontaneous sublattice moment develops is called the Néel temperature T_N. Solving of Eq. (4.3.9) leads to the expression $T_N = C'(N_1 \pm N_2)$, where

$$T_N = C'(N_1 - N_2) = \frac{1}{2} C(N_1 - N_2) \qquad (4.3.10)$$

is the correct solution. We know that $N_2 < 0$ and $N_1 > 0$. The solution $T_N = C'(N_1 + N_2)$ is not acceptable since, if $|N_2| > N_1$, this leads to a negative value of the magnetic-ordering temperature T_N, which is unphysical.

For temperatures above T_N, we may write

$$M = M_A + M_B = \frac{2C'}{T}\left(H + \frac{1}{2} N_1 M + \frac{1}{2} N_2 M\right) = \frac{C}{T}\left(H + \frac{1}{2} M(N_1 + N_2)\right). \qquad (4.3.11)$$

Since $\chi = M/H$, we find

$$\chi = \frac{C}{T - \frac{1}{2} C(N_1 + N_2)} = \frac{C}{T - \theta_p}, \qquad (4.3.12)$$

where the paramagnetic Curie temperature is now given by

$$\theta_p = \frac{1}{2} C(N_1 + N_2). \qquad (4.3.13)$$

It follows from Eq. (4.3.12) that the susceptibility of an antiferromagnetic material follows Curie–Weiss behavior, as in the ferromagnetic case. However, for antiferromagnets θ_p is not equal to the magnetic-ordering temperature ($\theta_p \neq T_N$).

If we compare Eq. (4.3.10) with Eq. (4.3.13), we conclude that θ_p is smaller than T_N, bearing in mind that N_2 is negative. In many types of antiferromagnetic materials, one has the situation that the absolute value of the intersublattice-molecular-field constant is larger than that of the intrasublattice-molecular-field constant. In these cases, one finds with Eq. (4.3.13) that θ_p is negative. The χ^{-1} plot displayed in Fig. 4.2.1d corresponds to this situation.

In a crystalline environment, frequently, one crystallographic direction is found in which the atomic magnetic moments have a lower energy than in other directions (see further Chapters 5 and 11). Such a direction is called the easy magnetization direction. When describing the temperature dependence of the magnetization or susceptibility at temperatures below T_N, we have to distinguish two separate cases, depending on whether the measuring field is applied parallel or perpendicular to the easy magnetization direction of the two sublattice moments. As can be seen from Fig. 4.3.2, the magnetic response in these two directions is strikingly different.

We will first consider the case where the field is applied parallel to the easy magnetization direction in an antiferromagnetic single crystal, with H parallel to the A-sublattice magnetization and antiparallel to the B-sublattice magnetization. The magnetization of both sublattices can be obtained by means of

$$M_{A,B,0}(T) = M_{A,B,0}(0) B_J(y_{A,B,0}), \qquad (4.3.14)$$

$$y_{A,B,0} = \frac{\frac{1}{2} g J \mu_B \mu_0 |H_{A,B,0}|}{kT}, \qquad (4.3.15)$$

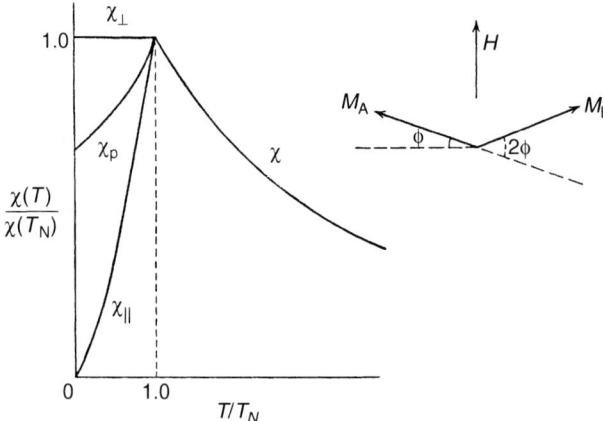

Figure 4.3.2. Temperature dependence of the magnetic susceptibility in an antiferromagnet with the field applied parallel and perpendicular to the easy magnetization direction of the two sublattices. Also shown is the bending of the two sublattice moments away from the easy direction for the case that the applied field is perpendicular to the easy direction.

SECTION 4.3. ANTIFERROMAGNETISM

where

$$H_A = H + N_1 M_A + N_2 M_B, \quad (4.3.16)$$

$$H_B = -H + N_1 M_B + N_2 M_A, \quad (4.3.17)$$

$$H_0 = N_1 M_B + N_2 M_A = \tfrac{1}{2}(N_1 + N_2)M. \quad (4.3.18)$$

Since the field is applied parallel to the A sublattice and antiparallel to the B sublattice, the A-sublattice magnetization will be slightly larger then the B-sublattice magnetization. The induced magnetization can then be obtained from $M = M_A - M_B = \Delta M_A + \Delta M_B$. For small applied fields, one may find ΔM_A and ΔM_B by expanding the corresponding Brillouin functions as a Taylor series in H and retaining only the first-order terms. After some tedious algebra, one eventually finds

$$\chi = \frac{\Delta M_A + \Delta M_B}{H} = \frac{N \mu_0 g^2 J(J+1) \mu_B^2 B'_J(y_0)}{kT - \tfrac{1}{2}(N_1 + N_2) N \mu_0 g^2 J(J+1) \mu_B^2 B'_J(y_0)}, \quad (4.3.19)$$

where $B'_J(y_0)$ is the derivative of the Brillouin function with respect to its argument. For more details, the reader is referred to the textbooks of Morrish (1965) and of Chikazumi and Charap (1966).

It can be inferred from Eq. (4.3.19) that $\chi = 0$ at zero kelvin and that χ increases with increasing temperature. The physical reason behind this is a very simple one. For both sublattices, the magnetically ordered state below T_N is due to the molecular field which leads to a strong splitting of the $2J + 1$ ground-state manifold (like in Fig. 3.1.1), so that in each of the two sublattices the statistical average value of $\langle \mu_z \rangle$ is nonzero when $H = 0$. The absolute values of $\langle \mu_z \rangle$ are the same for both sublattices, only the quantization directions of $\langle \mu_z \rangle$ are different because the molecular fields causing the splitting have opposite directions. If we now apply a magnetic field parallel to the easy direction, the total field will be slightly increased for one of the two sublattices, for the other sublattice it will be slightly decreased. This means that the total splitting of the former sublattice is slightly larger than in the latter sublattice. When calculating the thermal average $\langle \mu_z \rangle$ of both sublattices (Eq. 3.1.9), one finds that there is no difference at zero kelvin since for both sublattices only the lowest level is occupied and one has

$$M_A = M_B = \tfrac{1}{2} N \langle \mu_z \rangle = \tfrac{1}{2} N g J \mu_B \quad (4.3.20)$$

and consequently

$$\chi = \frac{\Delta M_A + \Delta M_B}{H} = 0. \quad (4.3.21)$$

However, as soon as the temperature is raised there will be thermal population of the $2J + 1$ levels. Because the total splitting for the two sublattices is different, one obtains different level occupations for both sublattices. The corresponding difference in the thermal averages $\langle \mu_z \rangle$ becomes stronger, the lower the population of the two lowest levels. In other words, although in both sublattices the statistical average $|\langle \mu_z \rangle|$ decreases with increasing temperature, the difference between $\langle \mu_z \rangle$ for the two sublattices increases and causes the susceptibility $\chi = (\Delta M_A + \Delta M_B)/H$ to increase with temperature (see Fig. 4.3.2).

We will now consider the susceptibility of an antiferromagnetic single crystal with the magnetic field applied perpendicular to the easy direction. The applied field will then produce a torque that will bend the two sublattice moments away from the easy direction, as is schematically shown in the inset of Fig. 4.3.2. This process is opposed by the molecular field that tries to keep the two sublattice moments antiparallel. The total torque on each sublattice moment must be zero when an equilibrium position is reached after application of the magnetic field. For the A-sublattice moment, this is expressed as follows:

$$\vec{\tau}_A = \mu_0 \vec{M}_A \times \vec{H}_A = 0, \quad (4.3.22)$$

with

$$\vec{H}_A = \vec{H} + N_1 \vec{M}_A + N_2 \vec{M}_B. \quad (4.3.23)$$

A similar expression applies to the torque $\vec{\tau}_B$ experienced by the B-sublattice moment but with $\vec{\tau}_B$ in a direction opposite to $\vec{\tau}_A$. Eq. (4.3.22) can be written as

$$\mu_0 \vec{M}_A \times \vec{H} + \mu_0 N_2 \vec{M}_A \times \vec{M}_B = 0$$
$$\mu_0 M_A H \cos \phi + \mu_0 N_2 M_A M_B \sin 2\phi = 0 \quad (4.3.24)$$
$$2 M_B \sin \phi = -\frac{H}{N_2}$$

The components of the two sublattice moments in the direction of the field lead to a net magnetization equal to

$$M_\perp = (M_A + M_B) \sin \phi = 2 M_B \sin \phi. \quad (4.3.25)$$

After combining Eqs. (4.3.24) and (4.3.25), one obtains

$$\chi_\perp = \frac{M_\perp}{H} = -\frac{1}{N_2}. \quad (4.3.26)$$

Since N_2 is negative, we may write

$$\chi_\perp = \frac{1}{|N_2|}. \quad (4.3.27)$$

This result shows that the susceptibility of an antiferromagnet measured perpendicular to the easy direction is temperature independent and that its magnitude can be used to determine the absolute value of the intersublattice-molecular-field constant.

If the applied field makes an arbitrary angle α with the easy direction, the susceptibility in the direction of the field, χ_α, can be calculated by decomposing the field into its parallel and perpendicular components:

$$M_\| = \chi_\| H_\| = \chi_\| H \cos \alpha, \quad (4.3.28)$$
$$M_\perp = \chi_\perp H_\perp = \chi_\perp H \sin \alpha. \quad (4.3.29)$$

The magnetization in the direction of the field is then given by

$$M_\alpha = M_\| \cos \alpha + M_\perp \sin \alpha = \chi_\| H \cos^2 \alpha + \chi_\perp H \sin^2 \alpha \quad (4.3.30)$$

SECTION 4.3. ANTIFERROMAGNETISM

and hence the susceptibility by

$$\chi_\alpha = \chi_\| \cos^2\alpha + \chi_\perp \sin^2\alpha. \qquad (4.3.31)$$

In a polycrystalline sample, one has crystallites with all orientations relative to the field. Since the number of orientations lying within $d\alpha$ of the inclination is proportional to $\sin\alpha\, d\alpha$, we have for the susceptibility of a piece of polycrystalline material or for a powder sample

$$\chi_{\text{poly}} = \int_0^{\pi/2} \chi_\alpha \sin\alpha\, d\alpha = \chi_\| \int_0^{\pi/2} \cos^2\alpha \sin\alpha\, d\alpha + \chi_\perp \int_0^{\pi/2} \sin^2\alpha \sin\alpha\, d\alpha \qquad (4.3.32)$$

This leads to

$$\chi_{\text{poly}} = \tfrac{1}{3}\chi_\| + \tfrac{2}{3}\chi_\perp, \qquad (4.3.33)$$

with

$$\chi_{\text{poly}} = \chi_\| = \chi_\perp \quad \text{at } T = T_N \qquad (4.3.34)$$

and

$$\chi_{\text{poly}} = \tfrac{2}{3}\chi_\perp \quad \text{at } T = 0. \qquad (4.3.35)$$

The above results for the magnetic susceptibilities are generally found to be in qualitative agreement with the properties observed for polycrystalline samples of several simple antiferromagnetic compounds. A sharp maximum in the susceptibility at the Néel temperature, or, equivalently, a sharp minimum in the reciprocal susceptibility, are generally considered as experimental evidence for the occurrence of antiferromagnetic ordering in a given material.

Let us consider the effect of an external field H on a magnetic material for which the magnetization is equal to zero before a magnetic field is applied. The work necessary to generate an infinitesimal magnetization is given by

$$\partial W = \mu_0 H \partial M.$$

The total work required to magnetize a unit volume of the material is

$$W = \int \mu_0 H\, dM. \qquad (4.3.36)$$

For antiferromagnetic materials and comparatively low magnetic fields, we may substitute $M = \chi H$ into this equation. After carrying out the integration, one finds for the free energy change of the system

$$\Delta F = -W = \tfrac{1}{2}\chi H^2. \qquad (4.3.37)$$

It can be seen in Fig. 4.3.2 that $\chi_\perp > \chi_\|$ below the Néel temperature T_N. This means that the application of a magnetic field to a single crystal of an antiferromagnetic material will always lead to a situation in which the two sublattice moments orient themselves perpendicular to the direction of the applied field or nearly so, as shown in the right part of Fig. 4.3.2. With increasing field strength, the bending of the two sublattice moments into the field direction becomes stronger until both sublattice moments are aligned parallel to the field direction and further increase of the total magnetization is no longer possible. The

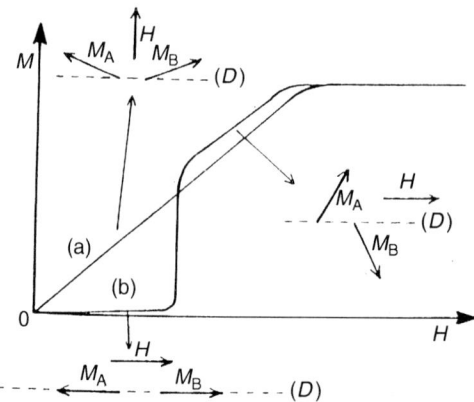

Figure 4.3.3. Schematic representation of the field dependence of the total magnetization in an antiferromagnetic single crystal in which the magnetocrystalline anisotropy is relatively low. The magnetic behavior is shown when measurements are made with the field applied in the hard direction (curve a) and in the easy direction (curve b). The easy direction is indicated by *D*. After Gignoux (1992).

field dependence of the magnetization behaves as shown by curve (a) in Fig. 4.3.3. The slope of the first part of this curve is given by $\Delta M/\Delta H = \chi_\perp$ and can be used to obtain an experimental value of the intersublattice-coupling constant N_2 according to Eq. 4.3.27.

In the discussion given above, we have assumed that the mutually antiparallel sublattice moments are free to orient themselves along any direction in the crystal. In other words, they can align themselves perpendicular to any direction in which the field is applied.

In most cases, however, the mutually antiparallel sublattice moments adopt a specific crystallographic direction in zero applied field. For this so-called easy direction, the magnetocrystalline anisotropy energy K (which will be discussed in more detail in Chapter 11) adopts its lowest value, $K = 0$. The field dependence of the magnetization will then show a behavior represented by curve (a) in Fig. 4.3.3 only if H is applied perpendicular to this easy direction.

Quite a different behavior will be observed when H is applied along the common easy direction of the two sublattice moments (indicated by D in Fig. 4.3.3). In this direction, the magnetocrystalline energy has its lowest value ($K = 0$), and the free energy is given by $\Delta F_\parallel = -\frac{1}{2}\chi_\parallel H^2$. By contrast, if the sublattice moments would adopt a direction perpendicular to the field direction and hence perpendicular to the easy direction (i.e., the so-called hard direction), the free energy would be given by $\Delta F_\perp = K - \frac{1}{2}\chi_\perp H^2$. For comparatively low applied fields, one has $\Delta F_\parallel < \Delta F_\perp$ and both sublattice moments will retain the easy moment direction. However, $\Delta F_\perp$ may become eventually the lowest energy state because $\chi_\perp > \chi_\parallel$. Both sublattice moments will therefore adopt a direction (almost) parallel to the applied field. The critical field H_c at which this happens is given by the equation

$$-\tfrac{1}{2}\chi_\parallel H_c^2 = K - \tfrac{1}{2}\chi_\perp H_c^2, \qquad (4.3.38)$$

which gives

$$H_c = \sqrt{\frac{2K}{\chi_\perp - \chi_\parallel}}. \quad (4.3.39)$$

This change in moment direction from the easy direction to a direction perpendicular to it is accompanied by an abrupt increase in the total magnetization, as illustrated by curve (b) in Fig. 4.3.3. This phenomenon is called spin flop. Of course, owing to the action of the field applied, the sublattice-moment directions are not strictly perpendicular to the easy direction. The sublattice moments have bent already into the field direction to some extent and will continue to do so above H_c for further increasing fields.

It is interesting to note that the magnetization corresponding to curve (b) for applied fields higher than H_c is slightly larger than that corresponding to curve (a). The reason for this is the following. The torque experienced by the sublattice moments due to the applied field that forces the sublattice moments into the field direction is counteracted in both cases by the intersublattice coupling that tries to keep the two sublattice moments mutually antiparallel (see previous section). In the case of curve (a), the torque produced by the applied field additionally has to overcome a restoring torque caused by the anisotropy energy that tries to keep the sublattice moments in the easy direction. This latter restoring torque acts in a favorable way in the case of curve (b) because the field is applied in the easy direction now. Therefore, for a given field strength above H_c, a larger degree of bending of the sublattice moments into the field direction is achieved in the case of curve (b) than in the case of curve (a).

A special situation is encountered in materials for which the magnetocrystalline anisotropy is very large. This is illustrated by means of Fig. 4.3.4 where the field dependence of the total magnetization M_t is plotted with the field applied in the hard direction (curve a) and in the easy direction (curve b). In the case of curve (a), the strong anisotropy prevents any sizable bending of the sublattice moments into the field direction. A forced

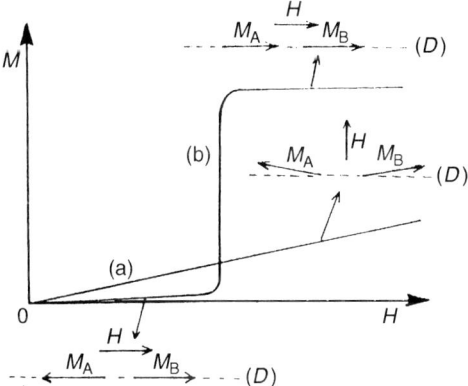

Figure 4.3.4 Schematic representation of the field dependence of the total magnetization in an antiferromagnetic single crystal in which the magnetocrystalline anisotropy is very strong. The magnetic behavior is shown when measurements are made with the field applied in the hard direction (curve a) and in the easy direction (curve b). The easy direction is indicated by D. After Gignoux (1992).

parallel arrangement of the two sublattice moments, as in the high-field part of curve (a) in Fig. 4.3.3, is not possible here. Therefore, the total magnetization remains low up to the highest field applied. In the case of curve (b), the total magnetization remains low for low fields. However, at a certain critical value of the applied field, the total magnetization jumps directly to the forced parallel configuration. We will compare now the free energy of the antiparallel sublattice-moment arrangement in the applied field with the parallel sublattice-moment arrangement in the applied field. Using Eqs. (4.3.1) and (4.3.2) for calculating $\Delta F = K - M_A H_A - M_B H_B$ for both situations and noting that $K = 0$ for all situations on curve (b), one easily derives the critical field as

$$H_c = N_{AB}(M_A + M_B) = N_2 M. \tag{4.3.40}$$

This formula expresses the fact that the sudden change from antiparallel to parallel sublattice-moment arrangement occurs when the applied field is able to overcome the antiferromagnetic coupling between the two sublattice moments. This phenomenon is called metamagnetic transition.

4.4. FERRIMAGNETISM

In ferrimagnetic substances, in contrast with the antiferromagnets described in the previous section, the magnetic moments of the A and B sublattices are not equal. The magnetic atoms (A and B) in a crystalline ferrimagnet occupy two kinds of lattice sites that have different crystallographic environments. Each of the sublattices is occupied by one of the magnetic species, with ferromagnetic (parallel) alignment between the moments residing on the same sublattice. There is antiferromagnetic (antiparallel) alignment, however, between the moments of A and B. Since the number of A and B atoms per unit cell are generally different, and/or since the values of the A and B moments are different, there is nonzero spontaneous magnetization below T_C. At zero Kelvin, it reaches the value

$$M_s = |N_A g_A J_A - N_B g_B J_B|\mu_B. \tag{4.4.1}$$

As in Eq. (4.1.2), we can represent the exchange interaction between the various spins S_i and S_j in the lattice by means of the Hamiltonian

$$H_{\text{exch}} = -\sum_{i<j} 2J_{ij} \vec{S}_i \cdot \vec{S}_j, \tag{4.4.2}$$

where J_{ij} is the exchange constant describing the magnetic coupling of two moments residing on the same magnetic sublattice A (or B) or on different sublattices A and B. Indicating the exchange constant between two nearest-neighbor spins on the same sublattice by J_{AA} (or J_{BB}) and between two nearest-neighbor spins on different sublattices by J_{AB}, we can represent the three types of cooperative magnetism leading to ordered magnetic moments as follows:

Ferromagnetism	$J_{AA} > 0$,		
Antiferromagnetism	$J_{AB} < 0$	and	$J_{AA} = J_{BB}$,
Ferrimagnetism	$J_{AB} < 0$	and	$J_{AA} \neq J_{BB}$.

SECTION 4.4. FERRIMAGNETISM

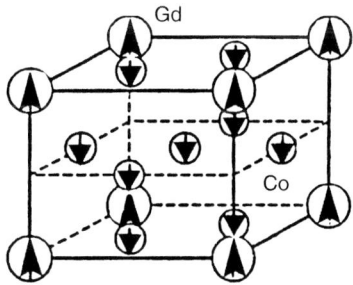

Figure 4.4.1. Arrangement of the magnetic moments in the unit cell of GdCo$_5$.

In general, J_{AA} and J_{BB} are positive quantities but this is not strictly necessary. For instance, there are ferrimagnetic Gd–Co compounds (see Fig. 4.4.1) in which $J_{GdCo} < 0$, $J_{CoCo} > 0$, and $J_{GdGd} < 0$, the strengths of these interactions decreasing in the sequence $|J_{CoCo}| > |J_{GdCo}| > |J_{GdGd}|$.

It will be shown in Chapters 12 and 13 that several of the most prominent magnetic materials are ferrimagnets. For this reason, we will discuss the magnetic coupling in these materials in somewhat more detail. We consider a ferrimagnetic compound consisting of two types of magnetic atoms A and B, occupying the sites of two different sublattices. The total angular moments of these magnetic atoms will be indicated as J_A and J_B. The corresponding g-factors are g_A and g_B, respectively. The magnetic moments per atom are related to the angular momenta by (Eq. 2.2.4):

$$\vec{\mu}_A = -g_A \mu_B \vec{J}_A \quad \text{and} \quad \vec{\mu}_B = -g_B \mu_B \vec{J}_B. \tag{4.4.3}$$

The exchange coupling between the various magnetic atoms can be described by means of Eq. (4.4.2). If we only take into account the magnetic interaction between the spins on nearest-neighbor atoms, the exchange interaction experienced by the spins S_A can be approximated by a molecular field H_A acting on J_A

$$H_{exch,A} = -2J_{AA} Z_{AA} \vec{S}_A \cdot \langle \vec{S}_A \rangle - 2J_{AB} Z_{AB} \vec{S}_A \cdot \langle \vec{S}_B \rangle - g_A \mu_B \vec{J}_A \vec{H}_A \tag{4.4.4}$$

A similar expression can be written down for the exchange interaction experienced by the spins S_B. The quantities J_{AA} (J_{BB}) and J_{AB} in Eq. (4.4.4) represent the exchange-coupling constants associated with the intrasublattice interaction and the intersublattice interaction, respectively. The number of similar neighbors and the number of dissimilar nearest neighbors are indicated as Z_{AA} (Z_{BB}) and Z_{AB}, respectively. From Eq. (4.4.4), we can derive an expression for the molecular field H_A by using $\langle \vec{S}_{A,B} \rangle = (g_{A,B} - 1)\langle \vec{J}_{A,B} \rangle$:

$$\vec{H}_A = -\frac{2J_{AA} Z_{AA}(g_A - 1)^2}{g_A \mu_B} \langle \vec{J}_A \rangle - \frac{2J_{AB} Z_{AB}(g_B - 1)(g_A - 1)}{g_A \mu_B} \langle \vec{J}_B \rangle \tag{4.4.5}$$

or, after using Eq. (4.4.3) and $\vec{M}_A = -N_A g_A \mu_B \langle \vec{J}_A \rangle$,

$$\vec{H}_A = \frac{2J_{AA} Z_{AA}(g_A - 1)^2}{N_A g_A^2 \mu_B^2} \vec{M}_A + \frac{2J_{AB} Z_{AB}(g_B - 1)(g_A - 1)}{N_A g_A g_B \mu_B^2} \vec{M}_B, \tag{4.4.6}$$

so that
$$\vec{H}_A = N_{AA}\vec{M}_A + N_{AB}\vec{M}_B, \quad (4.4.7)$$

where the intrasublattice- and intersublattice-molecular-field constants N_{AA} and N_{AB} are defined as

$$N_{AA} = \frac{2J_{AA}Z_{AA}(g_A - 1)^2}{N_A g_A^2 \mu_B^2}, \quad (4.4.8)$$

$$N_{AB} = \frac{2J_{AB}Z_{AB}(g_B - 1)(g_A - 1)}{N_A g_A g_B \mu_B^2}. \quad (4.4.9)$$

In the paramagnetic regime, in the presence of a magnetic field H, the two sublattice moments are given by

$$M_A = \frac{C_A}{T}(H + N_{AA}M_A + N_{AB}M_B), \quad (4.4.10)$$

$$M_B = \frac{C_B}{T}(H + N_{BB}M_B + N_{BA}M_A), \quad (4.4.11)$$

where

$$C_A = \frac{N_A g_A^2 \mu_B^2 J_A(J_A + 1)}{3k}. \quad (4.4.12)$$

N_A represents the number of A atoms per mole of atoms of the material. A similar expression holds for C_B. A solution of Eqs. (4.4.10) and (4.4.11) with $M_A \neq 0$, $M_B \neq 0$, and $H = 0$ can be found if

$$\begin{vmatrix} \frac{C_A N_{AA}}{T} - 1 & \frac{C_A N_{AB}}{T} \\ \frac{C_B N_{BA}}{T} & \frac{C_B N_{BB}}{T} - 1 \end{vmatrix} = 0. \quad (4.4.13)$$

The corresponding temperature, T_C, is now given by the relation

$$T_C = \tfrac{1}{2}(C_A N_{AA} + C_B N_{BB}) + \tfrac{1}{2}\sqrt{(C_A N_{AA} - C_B N_{BB})^2 + 4C_A C_B N_{AB} N_{BA}}, \quad (4.4.14)$$

where the various types of constants C and N are given by Eqs. (4.4.8), (4.4.9), and (4.4.12). For a given crystal structure, the number of nearest neighbors Z_{AA}, Z_{BB}, and Z_{AB} are known. In most cases, the values of g and J pertaining to the magnetic atoms are also known. Equation (4.4.14) then gives essentially a relation between the magnetic-ordering temperature and the magnetic-coupling constants J_{AA}, J_{AB}, and J_{BB}.

In deriving expressions for the total magnetization and sublattice magnetizations in the magnetically ordered regime, we will assume that the moments of the A and B sublattices are aligned strictly antiparallel. This is the case if N_{AB} is the only nonzero molecular-field constant or if N_{AB} is large compared to N_{AA} and N_{BB}. This assumption will be more carefully examined later. The sublattice moments are then given by

$$M_A(T) = M_A(0)B_{J,A}(x_A) \quad \text{and} \quad M_B(T) = M_B(0)B_{J,B}(x_B), \quad (4.4.15)$$

SECTION 4.4. FERRIMAGNETISM

where $B_{J,A}(x_A)$ and $B_{J,B}(x_B)$ are the Brillouin functions corresponding to the quantum numbers J_A and J_B, respectively, and where

$$M_A(0) = N_A g_A J_A \mu_B \quad \text{and} \quad M_B(0) = N_B g_B J_B \mu_B. \qquad (4.4.16)$$

It is to be noted that the two expressions for $M_A(T)$ and $M_B(T)$ in Eq. (4.4.15) are coupled equations since

$$x_A = \frac{g_A J_A \mu_B \mu_0}{kT}(N_{AA} M_A + N_{AB} M_B), \qquad (4.4.17)$$

$$x_B = \frac{g_B J_B \mu_B \mu_0}{kT}(N_{BB} M_B + N_{BA} M_A). \qquad (4.4.18)$$

The applied field H is assumed to be zero in Eqs. (4.4.17) and (4.4.18), since we are interested in the spontaneous moment $M_s(T)$. The temperature dependence of $M_s(T)$ can be derived from the expression

$$M_s(T) = |M_A(T) - M_B(T)|. \qquad (4.4.19)$$

Some illustrative examples of magnetization versus temperature curves are given in Figs. 4.4.2 and 4.4.3, where we have assumed that $|M_A(0)| < |M_B(0)|$. The situation shown in Fig. 4.4.2a refers to a compound in which the A-intrasublattice interaction is antiferromagnetic or only weakly ferromagnetic while the B-intrasublattice interaction is ferromagnetic and much stronger. As a result, the effective molecular field experienced by the A moments is smaller than that experienced by the B moments. This has as a consequence that $|M_A(T)|$ decreases more rapidly with temperature than $|M_B(T)|$. Figure 4.4.2b refers to a case where

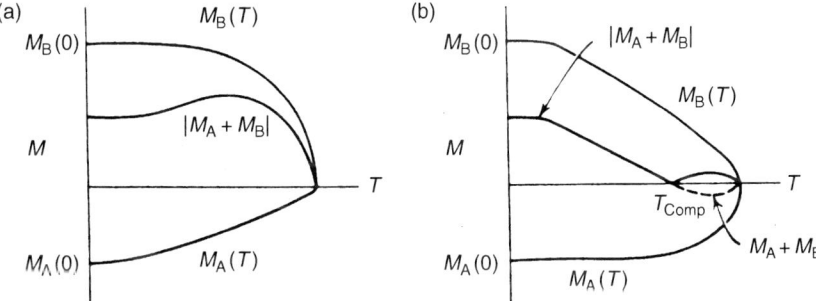

Figure 4.4.2. Two examples of spontaneous magnetization versus temperature curves predicted by the molecular field theory. (a) The net molecular field on the A-sublattice moments is smaller than on the B-sublattice moments, and (b) the molecular field is the strongest for the moments on the A sublattice.

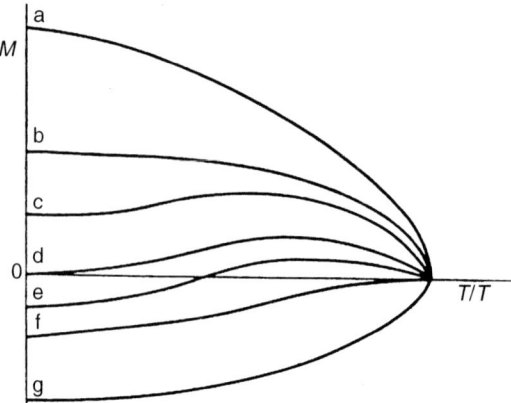

Figure 4.4.3. Resultant magnetization as a function of temperature for various values of $M_A(0)/M_B(0)$. Curves a, b, and c: $M_A(0)/M_B(0) < 1$. Curve b: $M_A(0)/M_B(0) = (N_{AB} - N_{AA})/(N_{AB} + N_{BB})$. Curve c: $M_A(0)/M_B(0) = (N_{AB} - N_{AA})/(N_{AB} - N_{BB})$. Curves e, f, and g: $M_A(0)/M_B(0) > 1$. After Gorter (1955).

the effective molecular field at the A sites is stronger than at the B sites. In this case, the spontaneous magnetization exhibits sign reversal. The temperature range in which this occurs is indicated by the dashed line. However, since the quantity measured in practice is $|M_A(T) + M_B(T)|$, the curve plotted as the full line is actually observed. The temperature T_{comp} at which the resultant magnetization is zero is commonly called the compensation point or compensation temperature.

Various other possible $M_s(T)$ curves are shown in Fig. 4.4.3. In practice, these different types of curves are observed when the composition of the compounds investigated is varied. For instance, there are various compounds in which rare earths (R) are combined with 3d metals (T), represented by the formula RT_n. There are several possibilities for choosing the T element (T = Ni, Co, Fe, Mn) and 15 possibilities for choosing the R element (see Table 2.2.1). An example of how the compensation temperature can be shifted to lower temperatures by reducing the R-sublattice magnetization via substitution of non-magnetic Y is shown in Fig. 4.4.4.

It follows from the discussion given above that the temperature dependence of the magnetization in ferrimagnetic compounds is determined by the magnitude and sign of the intrasublattice-coupling contants J_{AA} (J_{BB}) and the intersublattice-coupling constant J_{AB} appearing in Eqs. (4.4.8) and (4.4.9). If the sublattice moments $M_A(0)$ and $M_B(0)$ are known, these constants can be determined by fitting experimental curves of the temperature dependence of the total magnetization $M(T)$. The determination of three constants by fitting a simple $M(T)$ curve can, however, not always be accomplished in an unambiguous way. This is true, in particular when the $M(T)$ curve has not much structure. This is generally the case when it does not exhibit the singular point $M(T_{comp}) = 0$ at which the two sublattice moments become equal (Fig. 4.4.2b).

A most elegant and simple method, the high-field free-powder (HFFP) method, for determining the intersublattice-coupling constant has been provided by Verhoef et al. (1988). In this method, the molecular-field constant N_{RT} that determines the moment coupling

SECTION 4.4. FERRIMAGNETISM

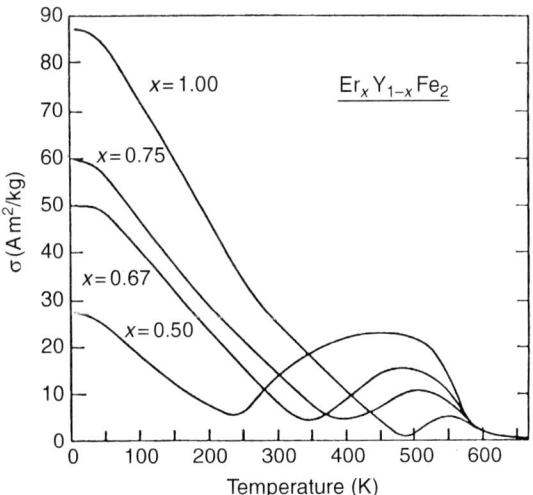

Figure 4.4.4. Temperature dependence of the magnetization of several $Er_xY_{1-x}Fe_2$ compounds measured in a field of 810 kA m^{-1}. After Buschow and van Stapele (1970).

between the rare-earth (R) sublattice and transition-metal (T) sublattice in ferrimagnetic intermetallic compounds is derived from magnetic measurements made on powder particles in high fields at low temperatures. The powder particles have to be sufficiently small in size so that they can be regarded as an assembly of small single crystals, able to rotate freely and orient their magnetization in the direction of the external field.

In many types of $R-T$ compounds, the anisotropy of the R sublattice exceeds that of the T sublattice by at least one order of magnitude at 4.2 K. By minimizing the free-energy, it can easily be shown that under such circumstances the low-temperature magnetization curve consists of three regions, as illustrated in Fig. 4.4.5. Below $H_{1,\text{crit}}$, there is a strictly antiparallel alignment between the (heavy)-R moments and the T moments, so that $M = |M_R - M_T|$. For sufficiently high values of the applied field, $H > H_{2,\text{crit}}$, the R and T moments are parallel and $M = M_R + M_T$. In the intermediate field range, $H_{1,\text{crit}} < H < H_{2,\text{crit}}$, there exists a canted-moment configuration, the R- and T-sublattice moments bending toward each other with increasing H. In this region, the field dependence of the total moment is given by

$$M = \mu_0 H / |N_{RT}|. \tag{4.4.20}$$

The slope of the $M(H)$ curve in the intermediate regime can therefore straightforwardly be used to determine the experimental value of N_{RT}, from which the coupling constant J_{RT} can be obtained via Eq. (4.4.9). A prerequisite for this method is that the two sublattice moments M_T and M_R do not differ too much in absolute value. The reason for this is that the first critical field

$$\mu_0 H_{1,\text{crit}} = N_{RT} |M_R - M_T| \tag{4.4.21}$$

has to be sufficiently low so that the linear magnetization region given by Eq. (4.4.20) falls within the experimentally accessible field range.

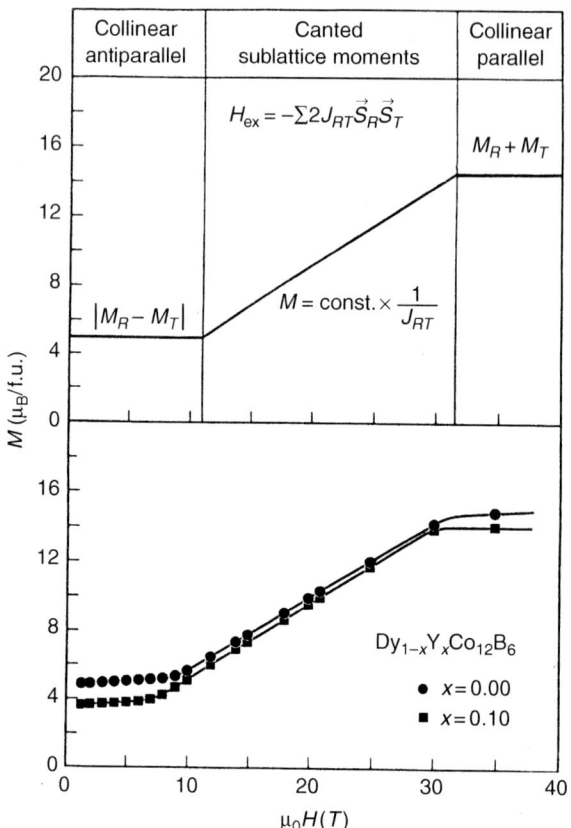

Figure 4.4.5. Schematic representation of the low-, medium-, and high-field behavior of magnetic isotherms expected on the basis of the HFFP method (top part). Experimental results obtained by the same method for two rare-earth cobalt borides of the $RCo_{12}B_6$ type. After Liu et al. (1994).

In general, it is found that J_{RT} is almost temperature independent. This means that reliable values of N_{RT} can also be derived in comparatively low fields for compounds having a compensation point in their temperature dependence of the magnetization. When measuring the field dependence of M at the latter temperature, one has $H_{1,\text{crit}} = 0$, and Eq. (4.4.20) applies already for low fields starting from the zero field. In fact, the presence at the compensation temperature of two antiparallel sublattice moments of equal size leads to a situation similar to that in an antiferromagnet below T_N. One could then equally well apply Eq. (4.3.27), where the intersublattice-molecular-field constant N_2 now takes the form N_{RT}. Magnetic dilution is another method to make the linear region given by Eq. (4.4.20) fall into the experimentally available field range. In such a case, the larger of the two sublattice magnetizations in Eq. (4.4.21) is reduced by substituting non-magnetic atoms for the magnetic atoms on this sublattice.

Inelastic neutron scattering is another method to determine intersublattice-coupling constants. This method is experimentally less easily accessible and will not be discussed

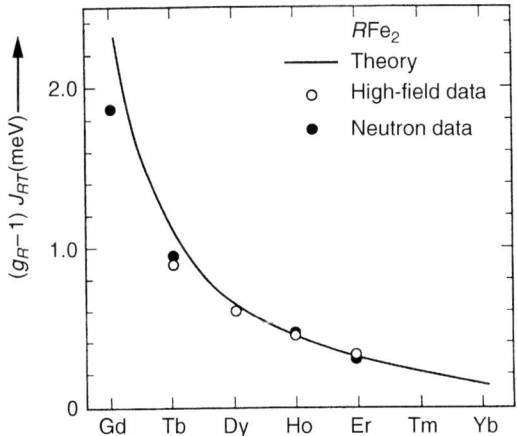

Figure 4.4.6. Intersublattice- magnetic coupling constants J_{RT} obtained by electronic band structure calculations made by Brooks and Johansson (1993) on RFe_2 intermetallics (full curve). Full symbols and open symbols represent experimental data obtained by inelastic neutron scattering and the HFFP method, respectively. After Liu et al. (1991).

here. Details of this method have been described by Nicklow et al. (1976) and Koon and Rhyne (1980). Results on RFe_2 compounds obtained by the HFFP method discussed above and results obtained by inelastic neutron scattering are compared with the results of electronic band structure calculations in Fig. 4.4.6. A compilation of intersublattic-coupling constants for various types of $R-T$ compounds has been presented by Liu et al. (1994).

References

Barbara, B., Gignoux, D., and Vettier, C. (1988) *Lectures on modern magnetism*, Beijing: Science Press.
Becker, R. and Döring, W. (1939) *Ferromagnetismus*, Berlin: Springer Verlag.
Beckman, O. and Lundgren, L. (1991) in K. H. J. Buschow (Ed.) *Handbook of magnetic materials*, Amsterdam. North Holland Publ. Co., Vol. 6, p. 181.
Brooks, M. S. S. and Johansson, B. (1993) in K. H. J. Buschow (Eds) *Handbook of magnetic materials*, Amsterdam: North Holland Publ. Co., Vol. 7, p. 139.
Buschow, K. H. J. and van Stapele, R. P. (1970) *J. Appl. Phys.*, 41, 4066.
Buschow, K. H. J. 1994 in R. W. Cahn et al. (Eds) *Materials science and technology*, Weinheim: VCH Verlag, Vol. 3B, p. 451.
Chikazumi, S. and Charap, S.H. (1966) *Physics of magnetism*, New York: John Wiley and Sons.
Gignoux, D. (1992) in R.W. Cahn et al. (Eds) *Material science and technology*, Weinheim: VCH Verlag, Vol. 3A, p. 267.
Gorter, E. W. (1955) *Proc. IRE*, 43, 1945.
Herring, C. (1966) in G. T. Rado and H. Suhl (Eds) *Magnetism*, New York: Academic Press, Vol. IIB, p. 1.
Koon, N. C. and Rhyne, J. J. (1980) in J. E. Crow et al. (Eds) *Crystalline electric fields and structure effects*, New York: Plenum, p. 125.
Liu, J. P., de Boer, F. R., and Buschow, K. H. J. (1991) *J. Magn. Magn. Mater*, 98, 291.
Liu, J. P., de Boer, F. R., de Châtel, P. F., Coehoorn, R., and Buschow, K. H. J. (1994) *J. Magn. Magn. Mater,* 132, 159.
Martin, D. H. (1967) *Magnetism in solids*, London: Iliffe Books Ltd.

Morrish, A. H. (1965) *The physical principles of magnetism*, New York: John Wiley and Sons.
Nicklow, R. M., Koon, N. C., Williams, C. M., and Milstein, J. B. (1976) *Phys. Rev. Lett., 36*, 532.
Slater, J. C. (1930) *Phys. Rev., 35*, 509; *Phys. Rev., 36*, 57.
Sommerfeld, A. and Bethe, H. (1933) in H. Geiger and K. Scheel (Eds) *Handbuch der physik*, Berlin: Springer, Vol. 24, Part 2, p. 595.
Verhoef, R., Quang, P. H., Franse, J. J. M., and Radwanski, R. J. (1988) *J. Magn. Magn. Mater., 75*, 319.
White, R. M. (1970) *Quantum theory of magnetism*, New York: McGraw-Hill.

5

Crystal Fields

5.1. INTRODUCTION

Almost all magnetic phenomena described in the preceding two chapters depend on the lifting of the degeneracy of the $(2J + 1)$-degenerate ground-state manifold by magnetic fields (internal and external) and on the occupation of the levels of this manifold as a function of magnetic-field strength and temperature.

Apart from magnetic fields, electrostatic fields are also able to lift the $(2J + 1)$-fold degeneracy. In order to see this, we will consider first the comparatively simple case of an atom with orbital angular momentum $L = 1$ situated in a uniaxial crystalline electric field of two positive ions located along the z-axis. In the free atom, the states $m_L = \pm 1, 0$ have identical energies and are degenerate. However, in the crystal lattice, the atom has a lower energy when the electronic charge cloud is close to the positive ions as in Fig. 5.1.1a than when it is oriented midway between the positive charges, as in Fig. 5.1.1b and c. The wave functions which give rise to these electronic charge densities have the form $zf(r)$, $xf(r)$, and $yf(r)$ and are called the p_z, p_x, and p_y orbitals, respectively. In the axially symmetric electric field considered in Fig. 5.1.1, the p_x and p_y orbitals are still degenerate. The three degenerate energy levels referred to the free atom are shown as a broken line in the right part of Fig. 5.1.1. Had the symmetry of the electric field been lower than axial, the degeneracy of the p_x and p_y orbitals would also have been lifted.

The crystalline electric field is able to orient the electronic charge cloud into an energetically favorable direction (situation a in Fig. 5.1.1). This means that the associated orbital moment also may have a preferred direction in the crystal. We have seen in Chapter 2 that the spin moment is tied to the orbital moment by means of the spin–orbit interaction. This implies that there also exists some directional preference for the spin moment.

In the next section, it will be shown how one can describe the effect of electrostatic fields by means of a quantum-mechanical treatment.

The reader who is more materials oriented will be mainly interested in the magnetic anisotropy resulting from the crystal–field interaction. This holds in particular for readers interested in rare earth based permanent-magnet materials. For these readers it is not strictly necessary to work through Sections 5.2–5.5. Instead, we offer in Section 5.6 a simple physical picture by means of which the magnetic anisotropy induced by the crystal field in

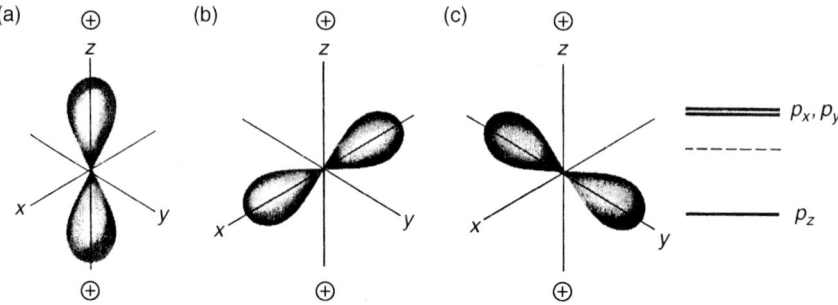

Figure 5.1.1. Three possible arrangements of the electronic charge cloud of an atom with orbital angular momentum $L = 1$ in the presence of an axially symmetric electrostatic field due to two positive charges. The right-hand part of the figure shows the effect of these positive charges on the three energy levels, the broken line representing the threefold-degenerate level in the absence of the charges. From Kittel (1968).

uniaxial rare-earth-based materials can be understood and by means of which the formulae used in Section 12.4 become sufficiently transparent.

5.2. QUANTUM-MECHANICAL TREATMENT

In most compounds, the magnetic atoms or ions form part of a crystalline lattice in which they are surrounded by other ions, the symmetry of the nearest-neighbor coordination being determined by the crystal structure. In ionic crystals, the metal ions are usually surrounded by negatively charged diamagnetic ions. Also in metallic systems, the constituting atoms carry an effective electric charge. This is due to the fact that they have donated all or at least a substantial part of their valence electrons to the conduction band. The resultant positive ions are screened to some extent by the conduction electrons, making the effective charge smaller than the corresponding ionic charges. The electrostatic field experienced by the unpaired electrons of a given magnetic ion is called crystal field or ligand field. The neighboring ions, surrounding the atom with the unpaired electrons, are called the ligands. A typical situation, where the atom carrying the unpaired electrons is situated in a uniaxial crystal field, is shown in Fig. 5.2.1.

If J is the total angular-momentum quantum number of the magnetic atom, the $(2J+1)$-fold degeneracy of its ground state will be lifted in the presence of a magnetic as well in the presence of a crystal field. This will result in changes in the magnetic properties of the corresponding compound if a crystal field is present.

In order to derive the magnetic properties, it is necessary to solve the Hamiltonian of the crystal–field interaction explicitly. The crystal-field potential due to the surrounding ions at the location of the kth unpaired electron of the magnetic ion, is

$$V_k(\vec{r}_k) = |e| \sum_j \frac{Z_j}{|\vec{R}_j - \vec{r}_k|}, \qquad (5.2.1)$$

where $|e|$ is the absolute value of the electron charge. The charge of the jth ligand ion is Z_j (Z_j can be either positive or negative). $\vec{R}_j$ and $\vec{r}_k$ are the positions of the jth ligand ion

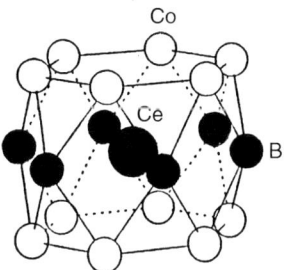

Figure 5.2.1. Nearest-neighbor configuration of Ce atoms in the intermetallic compound CeCo$_3$B$_2$ leading to a uniaxial crystal field. Uniaxial here means that the c-direction (vertical direction) is unique, whereas there are six equivalent directions perpendicular to c.

and the kth unpaired electron, respectively. The summation is carried out over all ligand ions in the crystal, taking the center of the magnetic ion considered as origin.

In a more rigorous treatment, the electric charges associated with the on-site valence electrons of the magnetic ion also have to be included in the crystal-field potential and the charges associated with the ligand atoms have to be included in the form of charge densities. The crystal-field potential then takes the form of an integration in space over all on-site and off-site charge densities around $\vec{r}_k$. We will return to this point later and use V_k as given above for introducing the operator equivalent method, without loss in generality. More rigorous treatments of crystal-field theory have been presented by Hutchings (1964), White (1970), and Barbara et al. (1988).

The crystal-field Hamiltonian of the magnetic ion is obtained from Eq. (5.2.1) by summing over all n_f unpaired electrons

$$H_{cf} = -|e| \sum_{k=1}^{n_f} V_k(\vec{r}_k). \tag{5.2.2}$$

The Hamiltonian may be expanded in spherical harmonics since the charges causing the crystal field are outside the shell of the unpaired electrons (4f electrons in the case of rare-earth atoms):

$$H_{cf} = \sum_{n=0}^{\infty} \sum_{m=-n}^{n} A_n^m \sum_{k=1}^{n_f} r_k^n Y_n^m(\theta_k, \varphi_k). \tag{5.2.3}$$

Here, A_n^m are the coefficients of this expansion. Their values depend on the crystal structure considered and determine the strength of the crystal–field interaction. For instance, if the point-charge model would be applicable, in which the ions of the crystal are described by point charges located at the various crystallographic positions, the coefficients A_n^m can be calculated by means of

$$A_n^m = -\frac{4\pi e^2}{2n+1} \sum_j \frac{Z_j}{R_j^{n+1}} Y_n^m(\theta_j, \varphi_j), \tag{5.2.4}$$

where the summation again extends over all ligand charges Z_j and the corresponding ligand positions $(R_j, \theta_j, \varphi_j)$ in the crystal. Without going into details about numerical computations of A_n^m in terms of point charges, we will keep the treatment general and consider them as numerical constants and focus our attention again on the Hamiltonian.

A relatively elegant form for this Hamiltonian can be obtained by using Stevens' Operator Equivalents method. First, the spherical harmonics $Y_n^m(\theta_k, \varphi_k)$ are expressed in Cartesian coordinates, $f(x, y, z)$, after which x, y, and z are replaced by J_x, J_y, and J_z, respectively. In this way, an operator is formed with the same transformation properties under rotation as the corresponding spherical harmonics. For instance

$$\sum_{k=1}^{n_f} Y_2^0 \rightarrow \sum_{k=1}^{n_f}(3z_k^2 - r_k^2) = \alpha_J \langle r^2 \rangle \left[3J_z^2 - J(J+1) \right] = \alpha_J \langle r^2 \rangle O_2^0, \quad (5.2.5)$$

$$\sum_{k=1}^{n_f} Y_2^2 \rightarrow \sum_{k=1}^{n_f}(x_k^2 - y_k^2) = \alpha_J \langle r^2 \rangle [J_x^2 - J_y^2] = \alpha_J \langle r^2 \rangle O_2^2, \quad (5.2.6)$$

where $\langle r^2 \rangle$ is the expectation value of the 4f radius, α_J is a constant (and where $J_x^2 - J_y^2$ may be replaced by $\frac{1}{2}[J_+^2 + J_-^2]$). Note that the introduction of the Operator Equivalents has the obvious advantage that the summation over n_f is no longer necessary. Equation (5.2.3) may now be rewritten as

$$H_{cf} = \sum_{n=0}^{\infty} \sum_{m=-n}^{n} A_n^m \Theta_n^m \langle r^n \rangle O_n^m = \sum_{n=0}^{\infty} \sum_{m=-n}^{n} B_n^m O_n^m. \quad (5.2.7)$$

For a magnetic ion with a given J value, the operator equivalents O_n^m are known.

A complete list of them and their relation to the spherical harmonics can be found in the paper by Hutchings (1964). The quantities Θ_n^m are so-called reduced matrix elements that do not depend on the azimuthal quantum number m (but depend on J). Values of these quantities are also listed in Hutchings' paper. The latter constants are frequently indicated by α_J, β_J, and γ_J for $n = 2, 4$, and 6, respectively.

Finally, it can be shown that for f electrons ($l = 3$), n cannot exceed 6 ($n \leq 6$). Furthermore, n must be even owing to inversion symmetry of the crystal-field potential. This means that the above summation (for f electrons) is effectively only over $n = 2, 4, 6$, since $n = 0$ gives an additive constant to the potential, which has no physical significance.

For crystal structures with uniaxial symmetry (tetragonal or hexagonal symmetry), it is sometimes sufficient to consider only the $n = 2$ terms and neglect the higher order terms. In this case, the crystal-field Hamiltonian takes the relatively simple form

$$H_{cf} = \alpha_J \langle r^2 \rangle \left[A_2^0 O_2^0 + A_2^2 O_2^2 \right] = B_2^0 O_2^0 + B_2^2 O_2^2. \quad (5.2.8)$$

In Table 5.2.1, an example of how the perturbation matrix may be obtained for the case $J = 5/2$ is given. In uniaxial systems, it is obvious to choose the c-axis as quantization axis or z-axis. The result is a lifting of the $(2J + 1 =)$ six fold degeneracy of the ground state. The perturbation leads to three doublet states that are linear combinations of the states $|\pm 5/2\rangle$, $|\pm 3/2\rangle$, and $|\pm 1/2\rangle$.

Table 5.2.1. Effect of a second-order crystal-field perturbation on the $(2J+1)$ manifold of $^2F_{5/2}(Ce^{3+})$

$$H_{cf} = B_2^0 O_2^0 + B_2^2 O_2^2 \tag{1}$$

$$O_2^0 = 3J_z^2 - J(J+1) \tag{2}$$

$$O_2^2 = J_x^2 - J_y^2 = \tfrac{1}{2}(J_+^2 + J_-^2) \tag{3}$$

$$J_+ = J_x + iJ_y \text{ (raising operator)} \tag{4}$$

$$J_- = J_x - iJ_y \text{ (lowering operator)} \tag{5}$$

$$\langle J, m | J_z | J, m \rangle = m \tag{6}$$

$$\langle J, m | J^2 | J, m \rangle = J(J+1) \tag{7}$$

$$\langle J, m+1 | J_+ | J, m \rangle = \sqrt{J(J+1) - m(m+1)} \tag{8}$$

$$\langle J, m-1 | J_- | J, m \rangle = \sqrt{J(J+1) - m(m-1)} \tag{9}$$

All other matrix elements vanish.

	$|5/2\rangle$	$|1/2\rangle$	$|-3/2\rangle$	$|-5/2\rangle$	$|-1/2\rangle$	$|3/2\rangle$
$\langle 5/2|$	$10B_2^0$	$\sqrt{10}B_2^2$	0	0	0	0
$\langle 1/2|$	$\sqrt{10}B_2^2$	$-8B_2^0$	$3\sqrt{2}B_2^2$	0	0	0
$\langle -3/2|$	0	$3\sqrt{2}B_2^2$	$-2B_2^0$	0	0	0
$\langle -5/2|$	0	0	0	$10B_2^0$	$\sqrt{10}B_2^2$	0
$\langle -1/2|$	0	0	0	$\sqrt{10}B_2^2$	$-8B_2^0$	$3\sqrt{2}B_2^2$
$\langle 3/2|$	0	0	0	0	$3\sqrt{2}B_2^2$	$2B_2^0$

(10)

Result: Three doublets of the type

$$\psi_p^+ = a_p|+5/2\rangle + b_p|+1/2\rangle + c_p|-3/2\rangle \tag{11}$$

$$\psi_p^- = a_p|-5/2\rangle + b_p|-1/2\rangle + c_p|+3/2\rangle \tag{12}$$

In several uniaxial crystal structures, symmetry causes the B_2^2 term to be absent. It may be easily checked in Table 5.2.1 that the perturbation matrix is then already diagonal in m. The three doublets are $|\pm 5/2\rangle$, $|\pm 3/2\rangle$, and $|\pm 1/2\rangle$.

All results described above follow from symmetry considerations. In order to obtain the relative energy positions of the three doublet levels, one has to know the sign and magnitude of B_2^0. The energy level scheme for $B_2^0 > 0$ is shown in Fig. 5.2.2. If the ligand charges are known accurately, the values of $B_2^0 = \alpha_J \langle r^2 \rangle A_2^0$ can be calculated by means of the point-charge model. It is possible, however, to consider B_2^0 as a parameter that can be determined experimentally. For instance, an experimental value for B_2^0 can be obtained if the magnetic susceptibility $\chi = \langle \mu_z \rangle_{av}/H$ is calculated by determining the thermal average of $\langle \mu_z \rangle$ over the crystal-field-split states for each temperature on the basis of Eq. (3.1.4), with B_2^0 and the concomitant level splitting as adjustable parameters. The calculated $\chi(T)$ curve is then fitted with the experimental $\chi(T)$ curve. Another relatively simple method to obtain an experimental value for B_2^0 consists in measuring the temperature dependence of the specific heat.

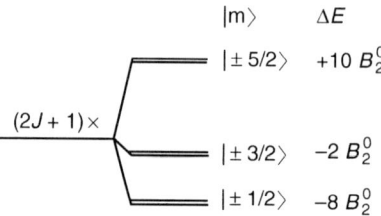

Figure 5.2.2. Effect of the electrostatic perturbation due to crystal fields on the $(2J+1)$-fold degenerate ground-state level for $J = \frac{5}{2}$.

Until now, we have used the 4f wave functions corresponding to the $|J, J_z\rangle$ representation to calculate the perturbing influence of the crystal field by means of the Hamiltonian given in Eq. (5.2.7). This means that we have tacitly assumed that the crystal–field interaction is small compared to the spin–orbit interaction introduced via the Russell–Saunders coupling and Hund's rules, and that J and m are good quantum numbers. Before applying this crystal-field Hamiltonian to 3d wave functions, we will first briefly review the relative magnitude of the energies involved in the formation of the electronic states. In the survey given below, we have listed the order of magnitude of the crystal-field splitting relative to the energies involved with the Coulomb interaction between electrons (as measured by the energy difference between terms), and the LS coupling in various groups of materials, comprising materials based on rare earths (R) and actinides (A). The numbers listed are given per centimeter.

Group	Shell	Term energy	LS coupling	Crystal-field splitting
Fe	3d	10^5	10	10^3
Pd, Pt	4d, 5d	10^4	10^2	10^4
R, A	4f, 5f	10^5	10^3	10^2

These energy values may be compared with the magnetic energy of a magnetic moment μ in a magnetic field B:

$$E_m = -\mu B.$$

Using typical values for μ ($1\mu_B$) and B (1T), one finds with $\mu_B = 0.9274 \times 10^{-23}$ A m^2 = 0.9274×10^{-23} J/T a magnetic energy equal in absolute value to 9.274×10^{-23} J or 4.670 cm^1. This then leads to the following sequences in energies:

For Fe-group materials: crystal field > LS coupling > applied magnetic field,

For rare-earth-based materials: LS coupling > crystal field > applied magnetic field.

The physical reason for this difference in behavior is the following: The 3d-electron-charge clouds reside more at the outside of the ions than the 4f-electron-charge clouds. Therefore, the former electrons experience a much stronger influence of the crystal field than the latter. The opposite is true for the spin–orbit interaction. This interaction is generally stronger,

SECTION 5.2. QUANTUM-MECHANICAL TREATMENT

the larger the atomic weight. Hence, it is larger for the rare earths than for the 3d transition elements.

In view of the energy consideration given above, one has to adopt the following procedure for dealing with these interactions. The spin–orbit interaction is the strongest interaction for rare-earth-based materials. Therefore, the spin–orbit coupling has to be dealt with first. Subsequently, the crystal-field interaction can be treated as perturbation to the spin–orbit interaction. This is how we have proceeded thus far, indeed. First, we have dealt with the spin–orbit interaction in the form of the Russell–Saunders coupling. The total angular momentum $\vec{J}$ and its J_z component are constants of the motion after application of the Russell–Saunders coupling, and J and m_J are good quantum numbers. Consequently, we have calculated the perturbing influence of the crystal field with the $|J, J_z\rangle$ representation as basis (see Table 5.2.1).

Table 5.2.2. Effect of an octahedral crystal field on the D term ($L = 2$)

Angular parts of the eigenfunctions

$$|2, -2\rangle = Y_2^{-2}, \ldots, |2, 2\rangle = Y_2^2$$

The nonzero matrix elements are

$$\langle 2, 0|Y_4^0|2, 0\rangle = 6$$
$$\langle 2, 1|Y_4^0|2, 1\rangle = \langle 2, -1|Y_4^0|2, -1\rangle = -4$$
$$\langle 2, 2|Y_4^0|2, 2\rangle = \langle 2, -2|Y_4^0|2, -2\rangle = 1$$
$$\langle 2, 2|Y_4^4|2, 2\rangle = \langle 2, -2|Y_4^0|2, -2\rangle = \sqrt{10}$$

	$|-2\rangle$	$|-1\rangle$	$|0\rangle$	$|1\rangle$	$|2\rangle$
$\langle -2|$	Dq	0	0	0	$5Dq$
$\langle -1|$	0	$-4Dq$	0	0	0
$\langle 0|$	0	0	$-4Dq$	0	0
$\langle 1|$	0	0	0	$-4Dq$	0
$\langle 2|$	$5Dq$	0	0	0	Dq

with $Dq \propto D_4 \langle r^4 \rangle$

Results after diagonalization:

One triplet at $-4Dq$:
$$\tfrac{1}{2}\sqrt{2}\left(Y_2^2 - Y_2^{-2}\right) \propto \frac{xy}{r}$$
$$\tfrac{1}{2}\sqrt{2}\left(Y_2^1 - Y_2^{-1}\right) \propto \frac{yz}{r}$$
$$\tfrac{1}{2}\sqrt{2}\left(Y_2^1 + Y_2^{-1}\right) \propto \frac{xz}{r}$$

One doublet at $+6Dq$:
$$\tfrac{1}{2}\sqrt{2}\left(Y_2^2 + Y_2^{-2}\right) \propto \frac{x^2 - y^2}{r^2}$$
$$Y_2^0 \propto \frac{3z^2 - r^2}{r^2}$$

In the case of 3d electrons, we have to proceed differently. First, we have to deal with the crystal–field interaction. Subsequently, we can introduce the spin–orbit interaction as a perturbation. Before application of the spin–orbit interaction, $\vec{L}$ and $\vec{S}$, and the corresponding z components L_z and S_z are constants of the motion and hence L, S, m_L, and m_S are good quantum numbers. Because the crystal–field interaction is of electrostatic origin, it affects only the orbital motion. Therefore, the crystal–field calculations can be made by leaving the electron spin out of consideration and using the wave functions $|L, L_z\rangle$ as basis set.

When calculating the matrix elements of the Hamiltonian given in Eq. (5.2.7), one has to bear in mind that only even values of n need to be retained. It can also be shown that terms with $n > 2l$ vanish ($l = 2$ for 3d electrons).

As an example, let us consider the crystal-field potential due to a sixfold cubic (or octahedral) coordination. Owing to the presence of fourfold-symmetry axes, only terms with $n = 4$ and $m = 0, \pm 4$ are retained, which leads to

$$H_{\text{oct}} = D_4 r^4 \left[\tfrac{7}{2} Y_4^0 + \tfrac{7\sqrt{10}}{4} \left(Y_4^4 + Y_4^{-4} \right) \right], \tag{5.2.9}$$

where the coefficients of the Y_n^m terms have been calculated with the help of Eq. (5.2.4), keeping D_4 as a constant depending on the ligand charges and distances. The calculations are summarized in Table 5.2.2 for a 3d ion with a D term as ground state.

If one calculates the expectation value of L_z for the various crystal-field-split eigenstates, one finds that $\langle L_z \rangle = 0$ for all of them. In other words, the crystal–field interaction has led to a quenching of the orbital magnetic moment. This is also the reason why the experimental effective moments in Table 2.2.2 are very close to the corresponding effective moments calculated on the basis of the spin moments of the various 3d ions.

5.3. EXPERIMENTAL DETERMINATION OF CRYSTAL-FIELD PARAMETERS

In order to assess the influence of crystal fields on the magnetic properties, let us consider again the situation of a simple uniaxial crystal field corresponding to a level splitting as in Fig. 5.2.2. If we wish to study the magnetization as a function of the field strength, we cannot use Eq. (3.1.9) because this result has been reached by a statistical average of μ_z based on an equidistant level scheme (see Fig. 3.1.1). Such a level scheme is not obtained when we apply a magnetic field to the situation shown in Fig. 5.2.2. The magnetic field will lift the degeneracy of each of the three doublet levels. Since a given magnetic field lowers and raises the energy of each of the sets of doublet levels in a different way, one may find a level scheme for $H \neq 0$ as shown in Fig. 5.3.1c. In order to calculate the magnetization, one then has to go back to Eq. (3.1.4).

Further increase of the applied field than in Fig. 5.3.1c would eventually bring the $|-5/2\rangle$ level further down to become the ground state, so that close to zero Kelvin one would obtain a moment of $\mu = \tfrac{5}{2} g \mu_B$. Again measuring at temperatures close to zero Kelvin, we would have obtained $\mu = \tfrac{1}{2} g \mu_B$ for applied fields much smaller than corresponding to Fig. 5.3.1c. This means that the field dependence of the magnetization at temperatures close to zero Kelvin looks like the curve shown in Fig. 5.3.2. The field required to reach $\tfrac{5}{2} g \mu_B$, and hence the shape of the curve, depends on the energy separation between the crystal-field

SECTION 5.3. EXPERIMENTAL DETERMINATION OF CRYSTAL-FIELD PARAMETERS

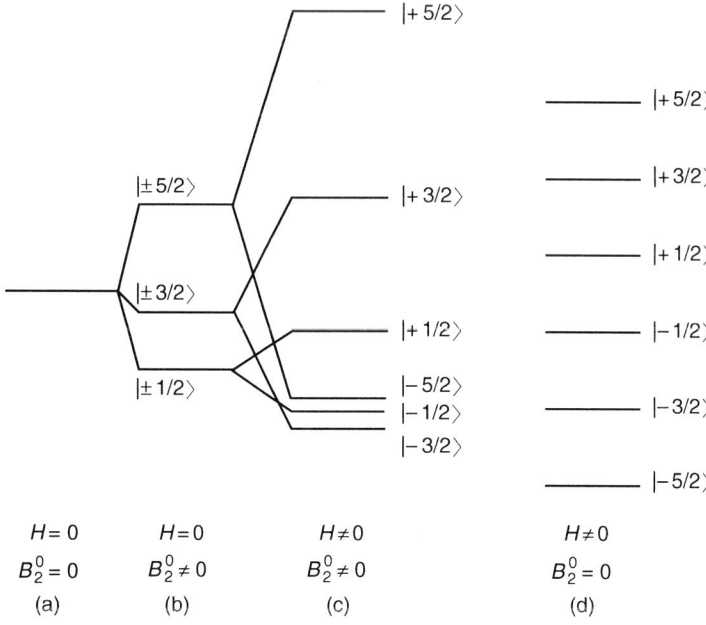

Figure 5.3.1. Level scheme obtained after splitting of the $(2J+1)$-fold-degenerate ground-state multiplet (a) by a uniaxial crystal field (b), by the simultaneous presence of the crystal field and a magnetic field applied in the uniaxial direction (c), and by the presence of only the magnetic field (d).

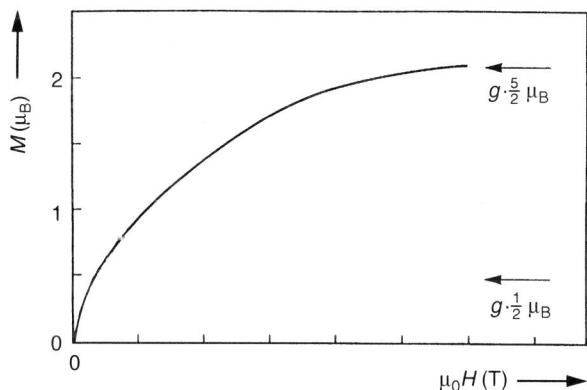

Figure 5.3.2. Schematic representation of the field dependence of the magnetization close to zero Kelvin for a crystal-field-split $J = \frac{5}{2}$ manifold, as shown in Fig. 5.3.1b.

split $|\pm 1/2\rangle$ and $|\pm 5/2\rangle$ levels. In other words, from a comparison of the measured $M(H)$ curve with curves calculated by means of Eq. (3.1.4) for various values of B_2^0, one may obtain an experimental value for the parameter B_2^0. Alternatively, one can keep H constant and vary the temperature. Subsequently, one can compare measured $M(T)$ or $\chi(T)$ curves with calculated curves (with B_2^0 again as adjustable parameter) and obtain in this way an experimental value of B_2^0. This procedure can also be followed in cases where more than one crystal-field parameter is required. In fact, it is just this process of curve fitting that reveals how many parameters are needed in each case and what their values are.

For completeness, we mention here that other experimental methods to determine sign and value of crystal-field parameters comprise inelastic neutron scattering and measurement of the temperature dependence of the specific heat. In the neutron-scattering experiment, the energy separation between the crystal-field-split levels of the ground-state multiplet is measured via the energy transfer during the scattering event between a neutron and the atom carrying the magnetic moment. In the specific-heat measurements, one obtains information on the change of the entropy with temperature. The entropy is given by $S = k \ln W$, where W is the number of available states of the system. Clearly, W can change substantially when more crystal-field levels become available by thermal population with increasing temperature. The way in which S changes with temperature, therefore, gives information on the multiplicity and energy separation of the crystal-field levels.

5.4. THE POINT-CHARGE APPROXIMATION AND ITS LIMITATIONS

Once the magnitudes (and signs) of the parameters B_n^m have been determined experimentally, one wishes, of course, to know the origin that causes the values of B_n^m to have a particular sign and magnitude in a given material. For simplicity, we will consider again the case of a simple uniaxial crystal field for which we have determined experimentally that $B_2^2 = 0$ and that it has a level scheme as shown in Fig. 5.2.2. Using Eq. (5.2.8), we have

$$B_2^0 = \alpha_J \langle r^2 \rangle A_2^0. \tag{5.4.1}$$

Since α_J is a constant for each rare-earth element with a given J value and since also the expectation values of the 4f radii $\langle r^2 \rangle$ are well-known quantities for all rare-earth elements, one may also say that the fitting procedure discussed above leads to an experimental value for the parameter A_2^0.

In Section 5.2, we mentioned already that the coefficients A_n^m, associated with the series expansion in spherical harmonics of the crystal-field Hamiltonian (Eq. 5.2.3), can be written in the point-charge model in the form of Eq. (5.2.4). In the particular case of A_2^0, after transformation into Cartesian coordinates, one has

$$A_2^0 = -\frac{|e|}{4} \sum_j \frac{Z_j(3z_j^2 - R_j^2)}{R_j^5}, \tag{5.4.2}$$

where the summation is taken over all ligand charges Z_j located at a distance $R_j(x_j, y_j, z_j)$ from the central atom considered. Since, in a given crystal structure, the distances between a given atom and its surrounding atoms are exactly known, it is possible to make *a priori* calculations of A_2^0 which then can be compared with the experimental value.

The main problem associated with this approximation is the assumption that the ligand ions can be considered as point charges. In most cases, the ligand ions have quite an extensive volume and the corresponding electrostatic field is not spherically symmetric. Also, the magnitude of Z_j, and in some cases even the sign of Z_j, is not accurately known.

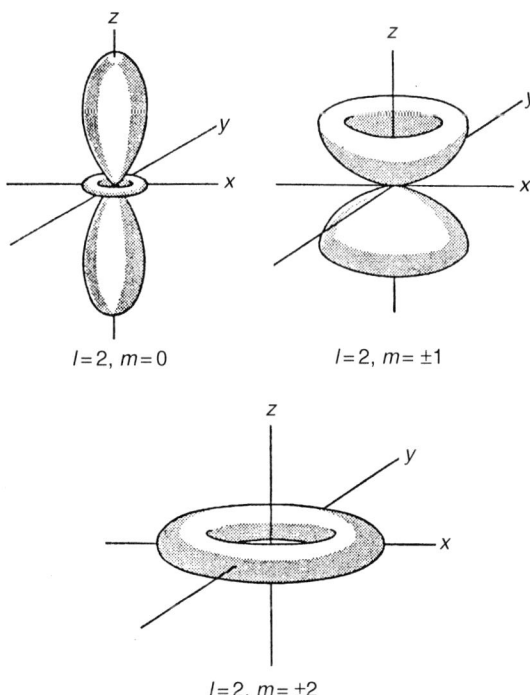

Figure 5.4.1. Orientation of charge clouds of d electrons ($l = 2$) in an axially symmetric electrostatic field.

The only benefit one may derive from the point-charge approximation is that it can be used to predict trends when crystal-field effects are compared within a series of compounds with similar structure.

A special complication exists in intermetallic compounds of rare-earth elements. This complication is due to the 5d and 6p valence electrons of the rare-earth elements. When placed in the crystal lattice of an intermetallic compound, the charge cloud associated with these valence electrons will no longer be spherically symmetric but may become strongly aspherical. This may be illustrated by means of Fig. 5.4.1, showing the orientations of d-electron-charge clouds with shapes appropriate for a uniaxial environment.

Depending on the nature of the ligand atoms, the energy levels corresponding to the different shapes in Fig. 5.4.1 will no longer be equally populated and produce an overall aspherical 5d-charge cloud surrounding the 4f-charge cloud. Similar arguments were already presented for p electrons in Fig. 5.1.1. Since the 5d and 6p valence electrons are located on the same atom as the 4f electrons, this on-site valence-electron asphericity produces an electrostatic field that may be much larger than that due to the charges of the considerably more remote ligand atoms. It is clear that results obtained by means of the point-charge approximation are not expected to be correct in these cases. Band-structure calculations made for several types of intermetallic compounds have confirmed the important role of the on-site valence-electron asphericities in determining the crystal field experienced by the 4f electrons (Coehoorn, 1992).

5.5. CRYSTAL-FIELD-INDUCED ANISOTROPY

As will be discussed in more detail in Chapter 11, in most of the magnetically ordered materials, the magnetization is not completely free to rotate but is linked to distinct crystallographic directions. These directions are called the easy magnetization directions or, equivalently, the preferred magnetization directions. Different compounds may have a different easy magnetization direction. In most cases, but not always, the easy magnetization direction coincides with one of the main crystallographic directions.

In this section, it will be shown that the presence of a crystal field can be one of the possible origins of the anisotropy of the energy as a function of the magnetization directions. In order to see this, we will consider again a uniaxial crystal structure and assume that the crystal–field interaction is sufficiently described by the $B_2^0 O_2^0$ term. Since we are discussing the situation in a magnetically ordered material, we also have to take into account a strong molecular field H_m as introduced in Section 4.1.

The energy of the system is then described by a Hamiltonian containing the interaction of a given magnetic atom with the crystal field and with the molecular field

$$H_{\text{tot}} = H_{\text{cf}} + H_{\text{exch}} = B_2^0 O_2^0 - \mu_0 \vec{\mu} \cdot \vec{H}_m. \tag{5.5.1}$$

The exchange interaction between the spin moments, as introduced in Eq. (4.1.2), is isotropic. This means that it leads to the same energy for all directions, provided that the participating moments are collinear (parallel in a ferromagnet and antiparallel in an antiferromagnet). So the exchange interaction itself does not impose any restriction on the direction of H_m. The two magnetic structures shown in Fig. 5.5.1 have the same energy when only the exchange term H_{exch} in the Hamiltonian is considered.

The examples shown in Fig. 5.5.1 are ferromagnetic structures ($J_{nn} > 0$) and the same reasoning can be held for antiferromagnetic structures ($J_{nn} < 0$) in which the moments are either parallel and antiparallel to c or parallel and antiparallel to a direction perpendicular to c. Also in these cases, the two antiferromagnetic structures have the same energy.

After inclusion of the $B_2^0 O_2^0$ term in the Hamiltonian, the energy becomes anisotropic with respect to the moment directions. This will be illustrated by means of the two ferromagnetic structures shown in Fig. 5.5.1. We assume that H_m is sufficiently large and

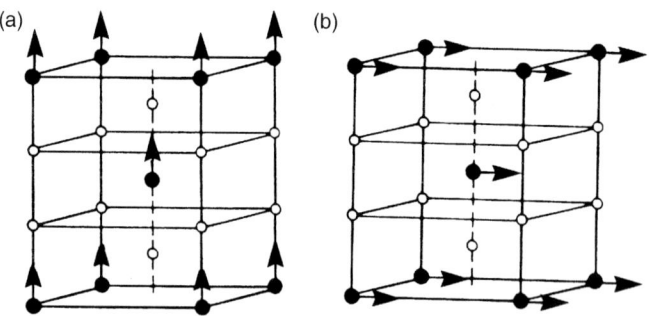

Figure 5.5.1. Moment arrangement in a ferromagnetic compound with the easy moment direction (a) along the c-axis or (b) along a direction perpendicular to the c-axis.

SECTION 5.5. CRYSTAL-FIELD-INDUCED ANISOTROPY

that the exchange splitting of the $|\pm 5/2\rangle$ level is much larger than the overall crystal-field splitting, $|-5/2\rangle$ being the ground state. The situation in Fig. 5.5.1a corresponds to $H_m//c$ or to $H_m//z$, since in crystal-field theory we have chosen the z-direction along the uniaxial direction. The situation in Fig. 5.5.1b corresponds to $H_m \perp c$, so that we may write $H_m//x$. Rewriting the Hamiltonian in Eq. (5.5.1) for both situations leads to

$$H_{\text{tot}} = B_2^0 \langle 3J_z^2 - J(J+1) \rangle + \mu_0 g \mu_B \langle J_z \rangle H_{m,z}, \quad (5.5.2)$$

$$H_{\text{tot}} = B_2^0 \langle 3J_z^2 - J(J+1) \rangle + \mu_0 g \mu_B \langle J_x \rangle H_{m,x}, \quad (5.5.3)$$

where

$$|H_{m,z}| = |H_{m,x}| = H_m = N_W M. \quad (5.5.4)$$

The Hamiltonian in Eq. (5.5.2) is already in diagonal form. Since we have chosen H_m large enough, the ground state is of course

$$|A\rangle = |-5/2\rangle.$$

One may easily obtain the ground-state energy by calculating

$$\langle A|H_{\text{tot}}|A\rangle = \langle -5/2|H_{\text{tot}}|-5/2\rangle = 10B_2^0 - \tfrac{5}{2}\mu_0 g \mu_B H_m.$$

In order to find the ground-state energy for $H_m//x$, one has to diagonalize the Hamiltonian in Eq. (5.5.3). This is a laborious procedure since the J_x operator will admix all states differing by $\Delta m = \pm 1$ ($J_x = \tfrac{1}{2}[J_+ + J_-]$, see Table 5.2.1). It can be shown that the ground-state wave function $|B\rangle$ is of the type

$$|B\rangle = a\,(|-5/2\rangle - |+5/2\rangle) + b\,(|-3/2\rangle - |+3/2\rangle) + c(|-1/2\rangle - |+1/2\rangle).$$

We will not further investigate this wave function except by stating that, owing to the predominance of H_m, it corresponds to an expectation value $\langle J_x \rangle = \langle B|J_x|B\rangle$ which is almost equal to $-\tfrac{5}{2}$. In fact, almost the full moment is obtained along the x-direction (at zero Kelvin). This means that the magnetic energy contribution is almost equal for the two cases (last terms of Eqs. 5.5.2 and 5.5.3).

On the other hand, one may notice that $\langle B|J_z^2|B\rangle = 0$, so that the crystal-field contribution in Eq. (5.5.3) is strongly reduced when the moments point into the x-direction. The energies associated with the Hamiltonians in Eqs. (5.5.2) and (5.5.3) can now be written as

$$E_z = 10B_2^0 - \tfrac{5}{2}\mu_0 g \mu_B H_m, \quad (5.5.5)$$

$$E_x = -5B_2^0 - \tfrac{5}{2}\mu_0 g \mu_B H_m. \quad (5.5.6)$$

It will be clear that E_z is lower than E_x for $B_2^0 < 0$. For $B_2^0 > 0$, the situation with the moments pointing along the x-direction is energetically favorable. These results can be summarized by saying that for a given crystal field ($B_2^0 < 0$ or $B_2^0 > 0$) the 4f-charge cloud adapts its orientation and shape in a way to minimize the electrostatic interaction with the crystal field. If the isotropic exchange fields experienced by the 4f moments are strong enough, one obtains the full moment $\mu = -g\mu_B m = g\mu_B J$ (or at least a value very close to it), but the direction of this moment depends on the sign of B_2^0.

5.6. A SIMPLIFIED VIEW OF 4f-ELECTRON ANISOTROPY

For the case of a simple uniaxial crystal field, we have derived in Section 5.2 that the leading term of the crystal–field interaction E_1 is given by the expectation value of $B_2^0 O_2^0$:

$$B_2^0 \langle O_2^0 \rangle = \alpha_J \langle r^2 \rangle A_2^0 \langle 3J_z^2 - J(J+1) \rangle. \tag{5.6.1}$$

In this section, we will show that the crystal–field interaction expressed in Eq. (5.6.1) can be looked upon in a different way, at the same time providing a simple physical picture for this type of crystal–field interaction. If the exchange interaction is much stronger than the crystal–field interaction, we showed in the previous section that ground state at zero Kelvin is $|J_z\rangle = |-J\rangle$. One then has

$$E_1 = \alpha_J \langle r^2 \rangle A_2^0 (2J^2 - J). \tag{5.6.2}$$

A_2^0 is the second-order term of Y_2^0 symmetry in the spherical harmonic expansion of the electrostatic crystal-field potential. This quantity can be looked upon as the gradient of the electric field.

Equation (5.6.1) then represents the interaction of the axial quadrupole moment associated with the 4f-charge cloud with the local electric-field gradient. It is good to bear in mind that a nonzero interaction with an electric quadrupole moment requires an electric-field gradient rather than an electric field.

The shape of the 4f-charge cloud resembles a discus if $\alpha_J < 0$. It resembles a rugby ball when $\alpha_J > 0$. Examples of both types of charge clouds are shown in Fig. 5.6.1.

It has already been mentioned that the molecular field in a magnetically ordered compound is isotropic and $H_m = N_W M$ has the same strength in any direction if the exchange coupling between the moments is the only interaction present. Alternatively, one may say that the magnetically ordered moments are free to rotate coherently into any direction. This directional freedom of the collinear system of moments is exploited by the interaction between the 4f-quadrupole moment and the electric-field gradient to minimize the energy expressed in Eq. (5.6.2). If the crystal field is comparatively weak, one may neglect any deformation of the 4f-charge cloud and the aspherical 4f-electron charge clouds shown in

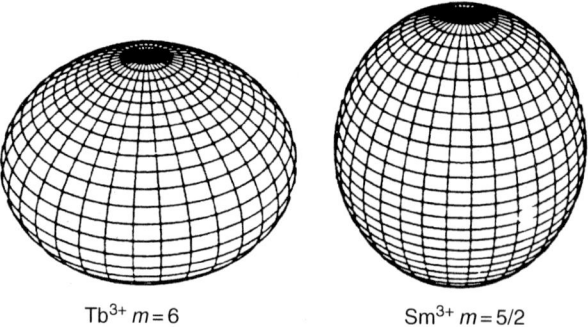

Tb^{3+} m=6   Sm^{3+} m=5/2

Figure 5.6.1. Angular distribution of the 4f-electron charge density of Tb^{3+} ($\alpha_J < 0$) and Sm^{3+} ($\alpha_J > 0$) for $m = -J$.

SECTION 5.6. A SIMPLIFIED VIEW OF 4f-ELECTRON ANISOTROPY

Table 5.6.1. Values of the second-order Stevens factor α_J and values of $\alpha_J \langle r^2 \rangle \langle O_2^0 \rangle$ at zero Kelvin (in units of a_0^2 with a_0 the Bohr radius) for several lanthanides

R^{3+}	$\alpha_J \times 10^2$	$\alpha_J \langle r^2 \rangle \langle O_2^0 \rangle$
Ce^{3+}	−5.714	−0.686
Pr^{3+}	−2.101	−0.713
Nd^{3+}	−0.643	−0.258
Sm^{3+}	+4.127	+0.398
Tb^{3+}	−1.010	−0.548
Dy^{3+}	−0.635	−0.521
Ho^{3+}	−0.222	−0.199
Er^{3+}	+0.254	+0.190
Tm^{3+}	+1.010	+0.454
Yb^{3+}	+3.175	+0.435

Fig. 5.5.2 will simply orient themselves in the field gradient to yield the minimum-energy situation.

It will be clear that for a crystal structure with a given magnitude and sign of A_2^0 the minimum-energy direction for the two types of shapes shown in Fig. 5.6.1 ($\alpha_J < 0$ and $\alpha_J > 0$) will be different. This implies that the preferred moment direction for rare-earth elements with $\alpha_J < 0$ and $\alpha_J > 0$ will also be different. It may be derived from Eq. (5.6.2) that the energy associated with preferred moment orientation in a given crystal field A_2^0 is proportional to $\alpha_J \langle r^2 \rangle (2J^2 - J)$. Values of this latter quantity for several lanthanides have been included in Table 5.6.1. A more detailed treatment of the crystal-field-induced anisotropy will be given in Chapter 12.

References

Barbara, B., Gignoux, D., and Vettier, C. (1988) *Lectures on modern magnetism*, Beijing: Science Press.
Coehoorn, R. (1992) in A. H. Cottrell and D. G. Pettifor (Eds) *Electron theory in alloy design*, London: The Institute of Materials, p. 234.
Hutchings, M. T. (1964) *Solid state phys.*, 16, 227.
Kittel, C. (1968) *Introduction to solid state physics*, New York: John Wiley & Sons.
White, R. M. (1970) *Quantum theory of magnetism*, New York: McGraw-Hill.

6

Diamagnetism

Diamagnetism can be regarded as originating from shielding currents induced by an applied field in the filled electron shells of ions. These currents are equivalent to an induced moment present on each of the atoms. The diamagnetism is a consequence of Lenz's law stating that if the magnetic flux enclosed by a current loop is changed by the application of a magnetic field, a current is induced in such a direction that the corresponding magnetic field opposes the applied field.

For obtaining expressions by means of which the diamagnetism of a sample can be described quantitatively, we will follow Martin (1967) and consider the perturbation of the orbital motion of electrons in the sample due to the force which each electron experiences when moving in a magnetic field. For a conductor element $\Delta \vec{l}$ carrying a current I in the presence of a magnetic field, this so-called Lorentz force is given by

$$\vec{F} = I \Delta \vec{l} \times \vec{B} \tag{6.1}$$

and hence in free space

$$\vec{F} = \mu_0 I \Delta \vec{l} \times \vec{H}. \tag{6.2}$$

If we consider the motion of a single charge e with velocity v we obtain

$$\vec{F} = \mu_0 e \vec{v} \times \vec{H}. \tag{6.3}$$

The effect of this force on an electron moving in a classical orbit around a single nucleus is easy to work out. It provides a picture that is not greatly changed in a quantum-mechanical treatment and is sufficient for our purpose. Let us assume that the field H is applied in a direction perpendicular to the plane of a circular orbit. The force F will act either away from the center of the orbit or toward it, depending on whether the electron is moving clockwise or anticlockwise with respect to the field. In either case, the change in the radius of the orbit can be neglected in comparison with the associated increase or decrease in the orbital angular velocity $\omega = v/r$. We will define the sign of ω as positive for clockwise orbital motion with respect to the field and negative for anticlockwise motion. Noting that the applied fields considered here are so small that they produce only small changes in v (and ω), and denoting such small incremental changes by Δ, we obtain, equating the magnetic force of Eq. (6.3) to mass times the change in acceleration,

$$e\mu_0 H \omega r = m\Delta(\omega^2 r) = 2mr\omega\Delta\omega$$

or

$$\Delta\omega = \frac{e\mu_0 H}{2m}. \tag{6.4}$$

The change in orbital angular velocity corresponds with a change in magnetic moment. If p represents the orbital angular momentum of the electron before application of the magnetic field, we may consider the equivalent magnetic shell and write

$$\mu = -\frac{|e|}{2m_e}p = -\frac{|e|}{2m_e}m_e r^2 \omega = -\frac{|e|}{2}r^2\omega. \tag{6.5}$$

The change in the magnetic orbital moment μ due to the field is

$$\Delta\mu = -\frac{|e|}{2}r^2 \Delta\omega = -\frac{\mu_0 e^2 r^2}{4m_e}H. \tag{6.6}$$

This equation shows that there is a negative change of the magnetic moment that is independent of the sign of ω and proportional to H.

If we consider a system consisting of N atoms, each containing i electrons with radii r_i, we may write for the susceptibility

$$\chi = \frac{\Delta\mu}{H} = -\frac{N\mu_0 e^2}{4m_e}\sum r_i^2. \tag{6.7}$$

In the derivation of this equation, we have assumed that the orbital plane of the electrons is perpendicular to the field direction. Instead of r_i^2 in Eq. (6.7), we should have used an effective radius q of the orbit such that $\langle q^2 \rangle = \langle x^2 \rangle + \langle y^2 \rangle$, $\langle q^2 \rangle$ representing the average of the square of the perpendicular distance of the electron from the field axis. The mean-square distance of the electrons from the nucleus is $\langle r^2 \rangle = \langle x^2 \rangle + \langle y^2 \rangle + \langle z^2 \rangle$, and since for a spherical symmetrical charge distribution one has $\langle x^2 \rangle = \langle y^2 \rangle = \langle z^2 \rangle$, one finds that $\langle q^2 \rangle = \frac{2}{3}\langle r^2 \rangle$. Using $\langle q_i^2 \rangle$ instead of r_i^2 in Eq. (6.7), leads to

$$\chi = -\frac{N\mu_0 e^2}{6m_e}\sum \langle r_i^2 \rangle, \tag{6.8}$$

which is the classical Langevin formula for diamagnetism.

In the quantum-mechanical treatment, one has to consider that the electrons are described by wave functions φ, where φ^2 at every point is the probability of finding the electron. Alternatively, one may consider the electron as a charge cloud of intensity φ^2 at each point in space. It can be shown that the quantum-mechanical result is correctly given by Eq. (6.8), provided one uses for $\langle r_i^2 \rangle$ the expectation value for the squared electron position parameter r_i:

$$\langle r_i^2 \rangle = \frac{\int r^2 |\varphi(r)|^2 \, dr}{\int |\varphi(r)|^2 \, dr}, \tag{6.9}$$

where the integration extends over the whole space.

We will close this chapter by mentioning that in metals there is a separate diamagnetic contribution due to the itinerant or band electrons to be discussed in Chapter 7.

If χ_{Pauli} represents the paramagnetic susceptibility due to these band electrons, it can be shown that it is accompanied by a diamagnetic contribution $\chi_{dia,band} = -\frac{1}{3}\chi_{Pauli}$.

Reference

Martin, D. H. (1967) *Magnetism in solids*, London: Iliffe Books Ltd.

7

Itinerant-Electron Magnetism

7.1. INTRODUCTION

A situation completely different from that of localized moments arises when the magnetic atoms form part of an alloy or an intermetallic compound. In these cases, the unpaired electrons responsible for the magnetic moment are no longer localized and accommodated in energy levels belonging exclusively to a given magnetic atom. Instead, the unpaired electrons are delocalized, the original atomic energy levels having broadened into narrow energy bands. The extent of this broadening depends on the interatomic separation between the atoms. According to a calculation made by Heine (1967), the following relation applies between the width of the energy bands W and the interatomic separation r:

$$W \propto r^{-5}. \tag{7.1.1}$$

The most prominent examples of itinerant-electron systems are metallic systems based on 3d transition elements, with the 3d electrons responsible for the magnetic properties. For a discussion of the magnetism of the 3d electron bands, we will make the simplifying assumption that these 3d bands are rectangular. This means that the density of electron states $N(E)$ remains constant over the whole energy range spanned by the bandwidth W.

A maximum of ten 3d electrons per atom (i.e., five electrons of either spin direction) can be accommodated in the 3d band. In the case of Cu metal, because each Cu atom provides ten 3d electrons, the 3d band will be completely filled. However, in the case of other 3d metals, less 3d electrons are available per atom so that the 3d band will be partially empty. Such a situation is shown in Fig. 7.1.1a. In Fig. 7.1.1a, we have indicated that there is no discrimination between electrons of spin-up and spin-down direction with respect to band filling. Both types of electrons will therefore be present in equal amounts, meaning that there is no magnetic moment associated with the 3d band in this case. However, this situation is not always a stable one, as will be discussed below.

It is possible to define an effective exchange energy U_{eff} per pair of 3d electrons. This can be regarded as the energy gained when switching from antiparallel to parallel spins. In order to realize such gain in energy, electrons have to be transferred, say, from the spin-down subband into the spin-up subband. As can be seen in Fig. 7.1.1b, this implies an increase in kinetic energy, which counteracts this electron transfer. However, it will be shown below that such transfer is likely to occur if U_{eff} is large and the density of states at the Fermi

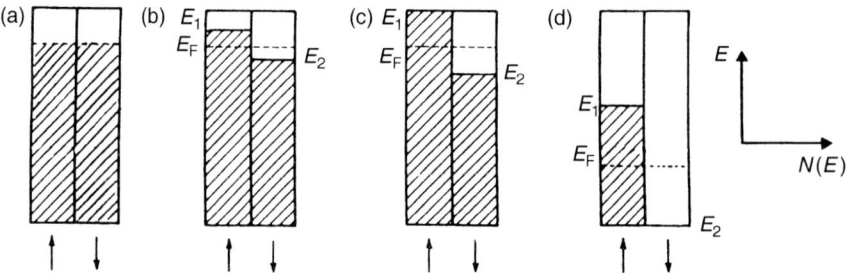

Figure 7.1.1. Schematic representation of a partially depleted 3d band: (a) paramagnetism, (b) weak ferromagnetism, (c) strong ferromagnetism with $n > 5$, (d) strong ferromagnetism with $n < 5$.

level E_F is high. After the transfer, there will be more spin-up electrons than spin-down electrons, and the magnetic moment, which has arisen, will be equal to $\mu = (n_1 - n_2)\mu_B$.

We will first derive a simple band model, which accounts for the existence of ferromagnetism. The interaction Hamiltonian, following the above definition of U_{eff}, can be written as

$$H = U_{\text{eff}} n_1 n_2, \quad (7.1.2)$$

where n_1 and n_2 represent the number of electrons per atom for each spin state, and where the total number of 3d electrons per atom equals $n = n_1 + n_2$. Because U_{eff} is a positive quantity, Eq. (7.1.2) will lead to the lowest energy if the product $n_1 n_2$ is as small as possible. For equally populated subbands, this product has its maximum value and hence the highest energy. Consequently, electron transfer is always favorable for the lowering of the exchange energy, and this electron transfer will come to an end only if one of the two spin subbands is empty or has become completely filled up.

We define $N(E)$ as the density of states per spin subband, and p as the fraction of electrons that has moved from the spin-down band to the spin-up band. This means

$$2pn = n_1 - n_2.$$

Let us assume that the interaction Hamiltonian (Eq. 7.1.2) leads to an increase in the number of spin-up electrons at the cost of the number of spin-down electrons. The corresponding gain in magnetic energy is then

$$\Delta E_M = U_{\text{eff}} n_1 n_2 - U_{\text{eff}} \tfrac{1}{4} n^2 = U_{\text{eff}} \tfrac{1}{2} n(1+2p) \tfrac{1}{2} n(1-2p) - U_{\text{eff}} \tfrac{1}{4} n^2,$$

$$\Delta E_M = -U_{\text{eff}} n^2 p^2.$$

This energy gain is accompanied by an energy loss in the form of the amount of energy ΔE_C needed to fill the states of higher kinetic energy in the band. For a small displacement $\delta E = E_F - E_2 = E_1 - E_F$ (see Fig. 7.1.1b), this kinetic-energy loss can be written as

$$\Delta E_C = \tfrac{1}{2} \delta E (n_1 - n_2) = \delta E n p.$$

The total energy variation is then

$$\Delta E_C + \Delta E_M = -U_{\text{eff}} n^2 p^2 + \delta E n p.$$

SECTION 7.2. SUSCEPTIBILITY ENHANCEMENT

Since
$$N(E_F)\delta E = \tfrac{1}{2}(n_1 - n_2),$$
one may write
$$\Delta E = \Delta E_C + \Delta E_M = \frac{n^2 p^2}{N(E_F)}[1 - U_{\text{eff}} N(E_F)]. \tag{7.1.3}$$

If $[1 - U_{\text{eff}} N(E_F)] > 0$, the state of lowest energy corresponds to $p = 0$ and the system is non-magnetic. However, if $[1 - U_{\text{eff}} N(E_F)] < 0$, the 3d band is exchange split ($p > 0$), which corresponds to ferromagnetism. The latter condition is the Stoner criterion for ferromagnetism, which is frequently stated in the more familiar form (Stoner, 1946)

$$U_{\text{eff}} N(E_F) > 1.$$

By means of this model, it can be understood that 3d magnetism leads to non-integral moment values if expressed in Bohr magnetons per 3d atom, $\mu = 2pn\mu_B$.

The conditions favoring 3d moments in metallic systems are obviously: a large value for U_{eff}, but also a large value for $N(E_F)$. The density of states of the s- and p-electron bands is considerably smaller than that of the d band, which explains why band magnetism is restricted to elements that have a partially empty d band. However, not all of the d-transition elements give rise to d-band moments. For instance, in the 4d metal Pd, the Stoner criterion is not met, although it comes very close to it.

7.2. SUSCEPTIBILITY ENHANCEMENT

The same formalism as used above can also be employed for calculating the magnetic susceptibility at zero temperature in a field H when the magnetic state is not stable with respect to the state without magnetic moment. The field will favor electron states with spin direction parallel to the field direction. If the latter is in the spin-up direction, the field will lead to a repopulation of the band states by transfer of electrons from the spin-down to the spin-up band. If p is the fraction of electrons transferred, we can use again Eq. (7.1.3) to calculate the energies involved in the electron transfer. Because a magnetic field is present, we have to add a Zeeman term $-2pn\mu_B\mu_0 H$ to the magnetic energy. This leads to

$$\Delta E = \Delta E_C + \Delta E_M = \frac{n^2 p^2}{N(E_F)}[1 - U_{\text{eff}} N(E_F)] - 2pn\mu_B\mu_0 H.$$

The equilibrium condition is
$$\frac{d\Delta E}{dp} = 0.$$

After differentiation of the expression for the energy, we find

$$\chi = \frac{2pn\mu_B}{H} = \frac{\chi_0}{1 - U_{\text{eff}} N(E_F)}, \tag{7.2.1}$$

where χ is the magnetic susceptibility per atom, and where χ_0 represents the "bare" unenhanced magnetic susceptibility which is given by

$$\chi_0 = 2\mu_B^2 N(E_F). \tag{7.2.2}$$

This quantity describes the magnetic susceptibility in metallic systems in which there is no interaction between the band electrons. This means that $U_{\text{eff}} = 0$ and thus that the Stoner enhancement factor $[1 - U_{\text{eff}} N(E_F)]^{-1}$ reduces to unity. By contrast, the enhancement factor can reach fairly high values when there is a strong interaction between the electrons and/or when the density of electron states at E_F is high. In fact, the Stoner enhancement factor can become extremely high for metallic systems close to magnetic instability, that is, when the Stoner criterion is fulfilled. We mentioned already that such a situation occurs for Pd metal. Experimentally, one finds that the susceptibility of Pd metal is roughly one order of magnitude larger than, for example, of Zr metal.

7.3. STRONG AND WEAK FERROMAGNETISM

In order to explain the principles of 3d-electron magnetism, up to now we have used a simplified model with rectangular 3d bands (Fig. 7.1.1). In a more realistic treatment, one has to take account of the actual shape of the 3d band in the energy range of interest (Friedel, 1969). This means that the density of states is no longer a constant over the whole energy range considered but may vary strongly as a function of the energy when moving from the bottom of the 3d band to the top. Generally, one finds that the density of states of the 3d band first increases when moving from the bottom of the 3d band in upward direction. After reaching a region where the density of states is high, one passes into a region where the density of states is fairly low. Moving further to the top of the band, one encounters again a region where the density of states is high.

When, for a given degree of 3d-band filling, the Fermi energy happens to be located in a region where the density of states is relatively low, the Stoner criterion may not be met and a spontaneous moment may form only if higher and lower lying states are included where the density of states is higher than in the immediate vicinity of the Fermi energy. Such a situation is schematically represented in Fig. 7.3.1.

Owing to the high kinetic-energy expenditure in the region where the density of states is low, no formation of a spontaneous moment will occur for small amounts of electron transfer, that is, for small 3d moments. The average energy expenditure is lower and spontaneous moments may form when electron states in the region with higher density of states are included. This implies that the formation of a spontaneous moment is only possible if the

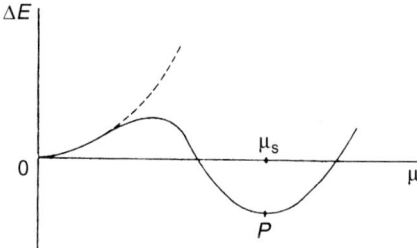

Figure 7.3.1. Energy dependence of magnetic moments (μ), showing the possibility of a stable magnetic phase when the paramagnetic phase is metastable. After Friedel (1969).

SECTION 7.3. STRONG AND WEAK FERROMAGNETISM

moments have a certain size. It will be shown below that the presence of a region in the 3d band with a low density of states can lead to two different situations.

Let us assume that the 3d band has a general shape of the type as indicated above, as is schematically shown in Fig. 7.3.1. In the case of simple ferromagnetism and relatively strong moments, it is possible to compute $\Delta E(\mu)$ directly from the expression:

$$\Delta E(\mu) = \int_{E_F}^{E_1} N(E) E \, dE + \int_{E_F}^{E_2} N(E) E \, dE - \frac{1}{4}(n_1 - n_2)^2 U_{\text{eff}}, \qquad (7.3.1)$$

where

$$n_i = \int_0^{E_i} N(E) \, dE. \qquad (7.3.2)$$

In analogy with the more simple case of rectangular bands, one can identify the first two terms in Eq. (7.3.1) as representing the loss in kinetic energy and the third term as representing the gain in exchange energy.

In order to have a ferromagnetic phase that is more stable than the paramagnetic phase, one has to meet the following condition:

$$\frac{d\Delta E(\mu)}{d\mu} \leq 0. \qquad (7.3.3)$$

When applied to Eq. (7.3.1), this means that a stable ferromagnetic state is found if

$$\frac{n_1 - n_2}{E_1 - E_2} = \frac{1}{E_1 - E_2} \int_{E_2}^{E_1} N(E) \, dE \geq \frac{1}{U_{\text{eff}}}. \qquad (7.3.4)$$

Generally, the maximum moment that can be obtained for a given number n of 3d electrons equals $(10-n)\mu_B$ for more than half-filled bands (see Fig. 7.1.1c) and $n\mu_B$ for less than half-filled bands (see Fig. 7.1.1d). However, when taking account of the kinetic-energy increase, the energy minimum of $\Delta E(\mu)$ in Fig. 7.3.1 may be reached for μ values considerably smaller than the maximum values of μ just mentioned (in analogy to the situation shown in Fig. 7.1.1b).

The situation shown for rectangular bands in Figs. 7.1.1b and d is shown for the more general band shapes in Figs. 7.3.2a and b, respectively. It is unphysical to have the Fermi energy at a higher level in the majority electron band than in the minority electron band. For this reason, the two subbands have been shifted relative to each other after electron transfer so as to have the same Fermi energy. This can also be interpreted by stating that the subband containing the larger number of electrons with parallel spins has been stabilized by the exchange energy with respect to the subband containing the lower number of electrons with parallel spins.

The situation in Fig. 7.3.2a corresponds to the equality sign in Eq. (7.3.4): For a given magnitude of U_{eff} the optimum band shift and the concomitant optimum electron transfer between the two subbands has been reached, the low density of states in the minority spin band preventing further electron transfer because of the too high kinetic energy expenditure. Note that both spin subbands remain partially depleted even though there are enough 3d electrons available for complete filling of the majority spin subband. The situation shown in Fig. 7.3.2a is referred to as weak ferromagnetism. The situation represented in

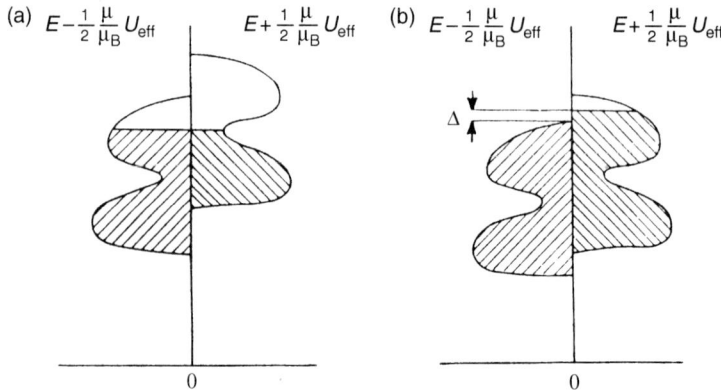

Figure 7.3.2. Relative position of the two half d-bands with opposite spin directions: (a) weak ferromagnetism, (b) strong ferromagnetism. After Friedel (1969).

Fig. 7.3.2b corresponds to the inequality sign in Eq. (7.3.4): The magnitude of U_{eff} and the corresponding band shift is larger than required for reaching the maximum moment for the degree of 3d-band filling considered in this figure. This situation is referred to as strong ferromagnetism. Note that the top of the majority-spin subband falls below the Fermi energy.

Which of these two types of ferromagnetism is reached in a given compound depends on the actual shape of the density of states curve, the total number of 3d electrons and the value of U_{eff}. The most interesting example is formed by the 3d metals themselves and their alloys. These systems usually have a bcc structure for which each of the two spin subbands is fairly well divided into two parts with a high density of states separated by a pronounced minimum in the density of states (as has been assumed in Fig. 7.3.2). It can be shown by means of Eq. (7.3.4) that for such a shape of $N(E)$ the depletion of the 3d band with decreasing number of 3d electrons proceeds as follows. Starting from a full 3d band, first one of the two spin subbands will become partially depleted (minority band) and this depletion continues until the upper portion of this subband is empty. This then leads to a further decrease of the number of 3d electrons to partial depletion of the other spin subband (majority band). This implies a simultaneous change from weak to strong ferromagnetism.

It is plausible that the increasing depletion of only the minority band in the regime of strong ferromagnetism leads to an increase of the magnetic moment with decreasing number of 3d electrons. This moment increase comes to an end, however, when the majority band also becomes more depleted. The reason for this can be described as follows. The Fermi level E_F in the majority band, the latter being only slightly depleted, is in a region of a high density of states. By contrast, the density of states at E_F in the minority band is at or close to the minimum in the density of states (as shown in the upper left part of Fig. 7.3.3). Consequently, when 3d electrons are further withdrawn from the 3d band, most of these electrons will come from the majority band where the density of states is high. This leads to a decreasing difference in the number of electrons of opposite spin direction, and hence to

SECTION 7.3. STRONG AND WEAK FERROMAGNETISM

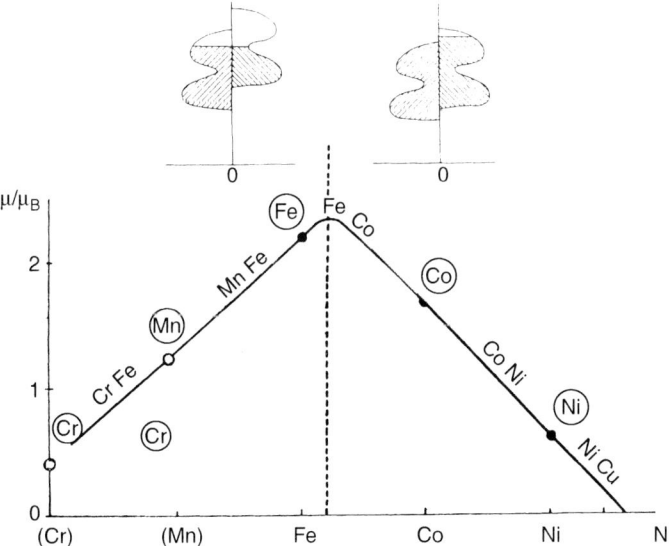

Figure 7.3.3. Slater–Pauling curve showing the variation of the mean 3d moment μ in 3d metals and their alloys as a function of the average atomic number (i.e., as a function of the number of 3d electrons N). After Friedel (1969).

a decrease in the 3d moment. This explains qualitatively the Slater–Pauling curve (Slater, 1937; Pauling, 1938) shown in the bottom part of Fig. 7.3.3. It is useful to bear in mind that the change from strong to weak ferromagnetism occurs close to Fe metal, which is a weak ferromagnet whereas the metals Co and Ni are strong ferromagnets. Other important points are

1. The designations strong ferromagnetism and weak ferromagnetism do not imply that the spontaneous moments per 3d atom or the magnetic ordering temperatures are higher in the former case than in the latter.
2. It has been shown in Section 4.2 that the magnetization in the fully ordered ferromagnetic state is given for localized moments by $M_s = Ng\mu_B J$. Once this state has been realized at low temperatures for a sufficiently high field, no further moment increase can be expected at still higher field strengths. The magnetization has become field-independent in a plot of M versus H and the high-field susceptibility defined in the saturated regime by $\chi = \Delta M/\Delta H$ is equal to zero. The reason for this behavior is the constancy of the localized moments. The situation is different, however, for itinerant moments. As we have seen above, the application of an external field stabilizes the majority-electron states with respect to the minority-electron states. This means that a small amount of electron transfer will be induced by a sufficiently high external field even in the saturated ferromagnetic state. Consequently, in a plot of M versus H the magnetization is not completely field independent and the high-field susceptibility defined in the saturated regime by $\chi = \Delta M/\Delta H$ is nonzero. Generally, the high-field susceptibility is larger for weak ferromagnets than for strong ferromagnets. Note that for the band shapes considered in Fig. 7.3.2 the high-field susceptibility for strong

ferromagnets is equal to zero because field-induced electron transfer into an already completely filled subband is not possible.

3. Many metal systems consist of a combination of a 3d transition metal (T) with a non-magnetic metal (A). Frequently, ferromagnetism disappears when the concentration of the T component becomes too low. This happens, for instance, in the series of intermetallic compounds formed by combining the non-magnetic element yttrium with cobalt: Y_2Co_{17}, YCo_5, Y_2Co_7, YCo_3, and YCo_2. The first four compounds are ferromagnetic with Curie temperatures much higher than room temperature, whereas the last compound does not show magnetic ordering at any temperature. It is wrong to say that the Co moment in the latter compound has disappeared because electron transfer from Y to the more electronegative Co has led to a filling up of the 3d band of the latter, preventing 3d magnetism. More realistic is the explanation that mixing of the Y valence-electron states with the Co 3d-electron states has led to a decrease of U_{eff} and to a broadening of the 3d band and a concomitant lowering of $N(E_F)$. The result is that 3d-band splitting will not occur, leaving the compound paramagnetic. Charge-transfer effects, where the valence electrons of A decrease the depletion of the 3d band of T do occur to some extent, but have a comparatively modest effect on the 3d-moment reduction upon alloying.

4. The application of the itinerant-electron model to the description of magnetism in 3d-electron systems does not necessarily mean that the 3d-electron spin polarization extends uniformly through the whole crystal. The small width of the 3d-electron band implies that the 3d electrons are rather strongly localized at the 3d atoms, and this holds a fortiori for their spin polarization. This justifies to some extent the use of local moments in molecular-field approximations for describing the magnetic coupling between 3d moments. It follows from the discussion given above that the moment of 3d atoms consists to a first approximation only of a spin moment. It is common practice to use the relation

$$\mu_{3d} = -g\mu_B S_{3d} \quad \text{with } g \approx 2. \tag{7.3.5}$$

7.4. INTERSUBLATTICE COUPLING IN ALLOYS OF RARE EARTHS AND 3d METALS

Metallic systems composed of magnetic rare-earth elements and magnetic 3d elements have found their way into many modern applications such as high-performance permanent magnets (Chapter 11), magneto-optic-recording materials (Chapter 13), and magneto-acoustic devices (Chapter 16). The favorable properties of these materials are partly due to the rare-earth sublattice (high magnetocrystalline anisotropy, high magnetostriction, high magnetic moments) and partly due to the 3d sublattice (high magnetic-ordering temperature). In order to have this combination of favorable properties in one and the same compound, it is of paramount importance that there be a strong magnetic coupling between the two magnetic sublattices involved.

There are several hundred intermetallic compounds composed of rare-earth metals and 3d metals and their magnetic properties are fairly well known and have been reviewed by Franse and Radwanski (1993). Without exception it is found that the rare-earth-spin moment couples antiparallel to the 3d-spin moment. This feature can be understood by means of an extension of the itinerant-electron model, as will be briefly described below. At first sight, an explanation in terms of the itinerant-electron model seems somewhat strange because we have treated 4f moments as strictly localized in Chapter 1. Also, in the present section we will deal with 4f electrons as localized, but additionally we will discuss the role played by the 5d valence electrons of the rare-earth elements. These 5d valence electrons are accommodated in narrow 5d bands, in a similar way as the 3d electrons of 3d transition metals are accommodated in 3d bands. In the rare-earth elements La and Lu, there are no 4f moments (see Table 2.2.1). From the magnetic properties of these elements it can be derived that the 5d electrons are not able to form 5d moments of their own. The reason for this is that the Stoner criterion (see Section 7.1) is not satisfied for the corresponding 5d bands. Nevertheless, these 5d electrons play a crucial role in the magnetic coupling between the 4f and 3d moments. Below, we will closely follow the treatment presented by Brooks and Johansson (1993).

Let us consider an isolated molecule of the compound LuFe$_2$. A schematic representation of the relative positions of the Lu 5d and Fe 3d atomic levels is shown in Fig. 7.4.1 before and after the two types of atoms have been combined to form a molecule. In the molecule, mixing of states leads to bonding and antibonding states, both states having a mixed 3d–5d character. Although this is of no particular concern in the present treatment, we will briefly mention that the electronic charges corresponding to the bonding states are accumulated mainly between the participating atoms. In the antibonding states, the electronic charges are accumulated mainly on the participating atoms. The bonding as well as the antibonding states broaden into bands when forming the solid compound, as illustrated in the left part of Fig. 7.4.2. Using the same simplified picture of rectangular bands as was done in the first part of Section 7.1, one can represent this situation by means of the diagram shown in the right part of Fig. 7.4.2.

Up to now, we have dealt equally with spin-up and spin-down electrons. However, from the fact that the Fe atoms carry a magnetic moment in LuFe$_2$, we know that the 3d band splits into a spin-up and a spin-down band, as discussed in the previous sections. This 3d-band splitting is illustrated in Fig. 7.4.3. The spin-up band, shown in the lower left part of the figure, is seen to be completely occupied. The spin-down band, shown in the lower right part is partly unoccupied. In fact, this difference in occupation reflects the presence of 3d moments on the Fe atoms.

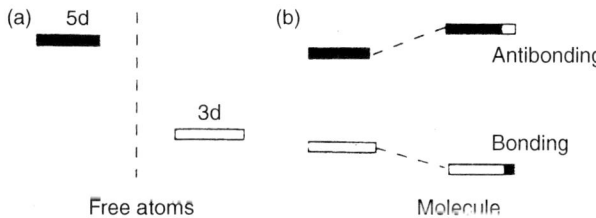

Figure 7.4.1. Schematic representation of the relative positions of the Lu 5d and Fe 3d atomic levels (a) before and (b) after the two types of atoms have been combined to a molecule. The black and white areas indicate the degree of mixing between the initial 3d and 5d states. After Brooks and Johansson (1993).

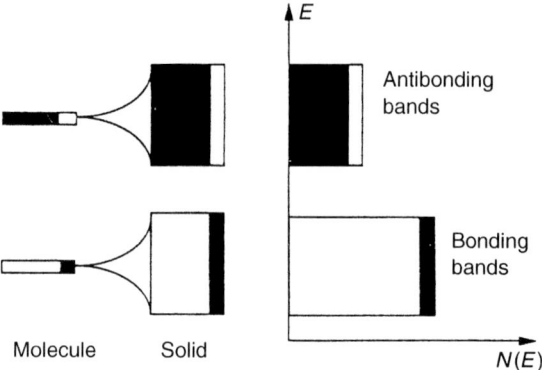

Figure 7.4.2. Schematic representation of the broadening of the molecular bonding and antibonding levels after condensation of LuFe$_2$ molecules to a solid (left). Diagram showing the corresponding density of states for rectangular d bands (right). The black and white areas indicate the degree of mixing between the initial 3d and 5d states. After Brooks and Johansson (1993).

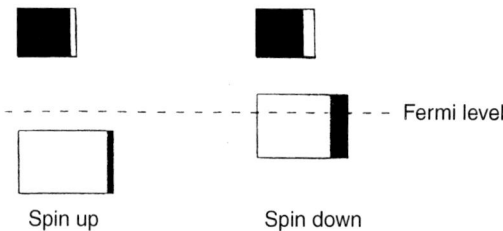

Figure 7.4.3. Schematic representation of the spin-up (majority) and spin-down (minority) density of states of the ferromagnetic compound LuFe$_2$. The black and white areas indicate the difference in the degree of mixing between the 3d- and 5d-band states for the two spin directions. After Brooks and Johansson (1993).

Closer inspection of the bands in Fig. 7.4.3 reveals the following. The spin-up and spin-down rare-earth 5d-band states mix with the transition-metal 3d-band states but do so to a different degree. The reason for this is the exchange splitting between the spin-up and spin-down 3d bands. For the spin-down electrons it leads to a smaller separation in energy between the 5d- and 3d-electron bands than for the spin-up electrons. Therefore, the mixing of 3d-band states and 5d-band states is larger for spin-up electrons than for spin-down electrons. As is indicated by the black and white areas in Fig. 7.4.3, the 5d(R)–3d(T) mixing leads to a larger 3d character of the 5d spin-down band than of the spin-up band. Consequently, the overall 5d(R) moment is antiparallel to the overall 3d(T) moment.

This mixing scheme is true for any rare-earth element R, independent of whether a 4f moment is present on the R atoms or not. When a 4f moment is present, one has ferromagnetic intra-atomic exchange interaction between the 4f-spin moment and the 5d-spin density, so that also the 4f-spin moment is antiparallel to the 3d moment. Summarizing these results, one can say that the rare-earth 5d electrons act as intermediaries for the coupling between the 3d and 4f spins which is always antiparallel. For more details, the reader is referred to the paper of Brooks and Johansson (1993).

Before closing this section, it is good to recall that the coupling scheme presented above is one between the spin moments. The itinerant model describes the 3d moments exclusively as spin moments, as mentioned already at the end of the previous section. We have seen in Section 7.2 that the 4f moments are composed of a spin moment and an orbital moment. For the heavy rare earths, we have $J = L + S$, meaning that the total 4f moment is also coupled antiparallel to the 3d moment. By contrast, we have $J = L - S$ for the light rare earths (see, for instance, Table 2.2.1). Consequently, the total 4f moment couples parallel to the 3d moment. In the two-sublattice model described in Section 4.4, with a negative spin–spin coupling ($J_{AB} = J_{RT} < 0$) for both cases, this different coupling behavior is taken account of by the different signs of the intersublattice-molecular-field constant $N_{AB} = N_{RT}$ in Eq. (4.4.9). It arises from the different signs of $g_R - 1$, being negative for the light rare earths but positive for the heavy rare earths (see Table 2.2.1).

References

Brooks, M. S. S. and Johansson, B. (1993) in K. H. J. Buschow (Ed.) *Handbook of Magnetic Materials*, Amsterdam: North Holland Publ. Co., Vol. 7, p. 139.
Franse, J. J. M. and Radwanski, R. J. (1993) in K. H. J. Buschow (Ed.) *Handbook of Magnetic Materials* Amsterdam: North Holland Publ. Co., Vol. 7, p. 307.
Friedel, J. (1969) in J. Ziman (Ed.) *The Physics of Metals*, Cambridge: Cambridge University Press, Vol. *1*, p. 340.
Heine, V. (1967) *Phys. Rev.*, *153*, 637.
Pauling, L. (1938) *Phys. Rev.*, *54*, 899.
Slater, J. C. (1937) *J. Appl. Phys.*, *8*, 385.
Stoner, E. C. (1946) *Rep. Progr. Phys.*, *9*, 43.

8

Some Basic Concepts and Units

Already in 1820, Ampère discovered that a magnetic field is produced by an electrical charge in motion. He showed that the magnetic field depends on the shape of the circuit and arrived at the result

$$I = \oint \vec{H} \cdot \vec{dl}, \tag{8.1}$$

which means that the current I in the conductor equals the line integral of H around an infinitely long rectilinear conductor. Performing the integration along a closed path around the conductor at a distance r leads to

$$\oint \vec{H} \cdot \vec{dl} = 2\pi r H = I \tag{8.2}$$

or

$$H = \frac{I}{2\pi r}. \tag{8.3}$$

In Chapter 6, we already introduced the force F experienced by a conductor element Δl carrying a current I in the presence of a magnetic field. In free space, Eq. (6.2) applies:

$$\vec{F} = \mu_0 I \vec{\Delta l} \times \vec{H}. \tag{8.4}$$

It can be easily shown from Eqs. (8.3) and (8.4) that if two infinitely long conductors (carrying currents I_1 and I_2) are mutually parallel and located at a distance d apart, the force per length exerted by one conductor on the other equals

$$\frac{F}{l} = \frac{\mu_0}{2\pi d} I_1 I_2. \tag{8.5}$$

Equation (8.5) is used to define the base unit of electric current, the ampere. The equation contains as a factor the magnetic permeability in vacuum or the magnetic constant μ_0. For historical reasons, this factor has been given the numerical value $4\pi \times 10^{-7}$. Using this, one arrives at the famous definition of the ampere: The ampere is that constant electric current which, if maintained in two straight parallel conductors of infinite length, of negligible circular cross-section and placed 1 m apart in vacuum, would produce between these conductors a force equal to 2×10^{-7} newton per meter of length. This definition implies, using Eq. (8.5), that the permeability in vacuum μ_0 takes the value

$$\mu_0 = 4\pi \times 10^{-7} \text{N A}^{-2}. \tag{8.6}$$

We define the unit of magnetic field strength in terms of the base unit ampere of electric current. For an infinitely long solenoid with n windings per length of solenoid, one finds inside the solenoid, by applying Ampere's law,

$$H = nI. \tag{8.7}$$

As n is expressed in m^{-1} and the electric current I in A, the magnetic field H has the unit ampere per meter or A m^{-1}. Consequently, a magnetic field of 1 A m^{-1} is the magnetic field in an infinitely long solenoid consisting of n turns per meter of coil and carrying an electric current of n^{-1} A.

The unit of the magnetic induction or magnetic flux density B can be defined by rewriting Eq. (8.4) as

$$\vec{F} = I \vec{\Delta l} \times \vec{B}. \tag{8.8}$$

This equation defines the magnetic induction B for any medium such that the force exerted on a current element $I \Delta l$ is equal to the vector product of this element and the magnetic induction. As the current element is expressed in Am and the force in N, the unit of magnetic induction is newton per ampere meter or $\text{N A}^{-1} \text{m}^{-1}$, which is called tesla, that is, $1 \text{ T} = 1 \text{ N A}^{-1} \text{m}^{-1}$. So the magnetic induction is equal to 1 T if a current element of 1 A experiences a force of 1 N. As the magnetic permeability in vacuum μ_0 equals $4\pi \times 10^{-7} \text{ N A}^{-2}$, a magnetic field strength of 1 A m^{-1} in free space corresponds to a magnetic induction of $4\pi \times 10^{-7}$ T or, equivalently, a magnetic induction of 1 T corresponds to a magnetic field of approximately 800 kA m^{-1}.

The magnetic flux $\Delta \Phi$ through a surface element ΔA is the scalar product of the magnetic flux density and this surface element:

$$\Delta \Phi = B \Delta A. \tag{8.9}$$

The unit of magnetic flux density B is tesla and so the unit of magnetic flux Φ is equal to tesla square meter or T m^2, which is called weber (Wb), that is $1 \text{ Wb} = 1 \text{ T m}^2 = 1 \text{ N m A}^{-1}$ or also $1 \text{ T} = 1 \text{ Wb m}^{-2}$.

Another definition of the magnetic flux comes from the phenomenon of induction: its rate of change $d\Phi/dt$ generates an electromotive force in a closed conductor. If the current passes through n turns of the conductor, the e.m.f. is given by:

$$E = -n \frac{d\Phi}{dt}. \tag{8.10}$$

This equation can also be used to define the unit of flux. As the electromagnetic force is expressed in volt and n is a pure number, the unit of magnetic flux becomes volt second or V s. Remembering that the unit of energy, joule, can be written as $1 \text{ J} = 1 \text{ V A s} = 1 \text{ N m}$, we see that the unit of magnetic flux can be transformed into $1 \text{ V s} = 1 \text{ N m A}^{-1}$ which is again equal to 1 Wb.

CHAPTER 8. SOME BASIC CONCEPTS AND UNITS

The magnetic induction B is connected with the magnetic field strength H by the equation

$$B = \mu H. \tag{8.11}$$

where μ is the magnetic permeability. As the magnetic induction is expressed in $N\,A^{-1}m^{-1}$ and the magnetic field in $A\,m^{-1}$, the permeability μ has, similar to the permeability in vacuum μ_0, the unit $N\,A^{-2}$, which is also called henry per meter, that is, $1\,H\,m^{-1} = 1\,N\,A^{-2}$. The relative permeability $\mu_r = \mu/\mu_0$ is of course a dimensionless number.

The simplest circuit which is able to generate a magnetic field is a planar circular conductor carrying an electric current. In fact, such a current loop can be considered as the most elementary unit of magnetism, as we saw already in Chapter 2. If a current loop has an area A and carries a current I, the corresponding magnetic moment is $\mu = IA$. The unit of magnetic moment can therefore be taken as ampere square meter $= A\,m^2$. When expressed in these units, the Bohr magneton introduced in Eq. (2.1.2) has a magnitude of $\mu_B = 0.9274 \times 10^{-23}\,A\,m^2$.

The magnetization can be defined as the magnetic moment per volume

$$M = p\mu, \tag{8.12}$$

where p is the number of moments μ per volume. With μ in $A\,m^2$ and p in m^{-3}, this gives $A\,m^{-1}$ for the unit of magnetization. When M is defined in this manner, one has the same units for the magnetization and the magnetic field strength.

In practice, it is more appropriate to define the magnetization as the magnetic moment per mass. The magnetization σ is then defined by

$$\sigma = p'\mu, \tag{8.13}$$

where p' is the number of magnetic moments per mass. With μ in $A\,m^2$ and p' in kg^{-1} this leads to $A\,m^2\,kg^{-1}$ for the unit of magnetization. The advantage of this choice of unit is that we do not need to know the volume of the sample of which we wish to determine the magnetization but only its mass, the latter being easily obtained by weighing the sample.

For comparison let us introduce the magnetic moment of N_A Bohr magnetons, where N_A is Avogadro's constant: 6.022×10^{23} molecules (or formula units) which are contained in one mole of material:

$$N_A\mu_B = (6.022 \times 10^{23}\,\text{mol}^{-1})(0.9274 \times 10^{-23}\,A\,m^2) = 5.585\,A\,m^2\,\text{mol}^{-1}. \tag{8.14}$$

We have the following relation between the various types of magnetizations

$$\frac{\mu}{\mu_B} = \frac{M_A\sigma}{N_A\mu_B} = \frac{M_A M}{N_A\mu_B\rho}, \tag{8.15}$$

where M_A represents the molar mass of the material, expressed in $kg\,mol^{-1}$. Introducing everywhere numerical values $\{..\}$ by substituting $M_A = \{M_A\}\,kg\,mol^{-1}$, $\sigma = \{\sigma\}\,A\,m^2\,kg^{-1}$, $M = \{M\}\,A\,m^{-1}$, $\rho = \{\rho\}\,g\,m^{-3}$, $N_A\mu_B = 5.585\,A\,m^2\,mol^{-1}$ and crossing out the units in the numerator and denominator, we obtain from Eq. (8.15) the equation between numerical values:

$$\frac{\mu}{\mu_B} = \frac{\{M_A\}\{\sigma\}}{5.585} = \frac{\{M_A\}\{M\}}{5.585\{\rho\}}. \tag{8.16}$$

We will now turn to the magnetic susceptibility, defined as $\chi = M/H$ with H in $A\,m^{-1}$. In accordance with the definition of M as expressed in Eq. (8.12), we may define the volume susceptibility which is a dimensionless quantity since M and H are both expressed in $A\,m^{-1}$. Based on Eq. (8.13), we can define the mass susceptibility $\chi_m = \sigma/H$ which has the dimension $(A\,m^2\,kg^{-1})/(A\,m^{-1}) = m^3\,kg^{-1}$. Division of the mass susceptibility by the molar mass leads to the molar susceptibility χ_M with unit $m^3\,mol^{-1}$.

It follows from the results discussed in the preceding chapters that if a material is placed in an external magnetic induction, B_0, different types of magnetic behavior can be observed, comprising diamagnetism, paramagnetism, or ferromagnetism. It will be clear that in diamagnetic materials, the internal magnetic induction, B_{int}, is somewhat smaller than the external magnetic induction, B_0. By contrast, in a paramagnetic material, the internal magnetic induction is somewhat larger than the external magnetic induction. In a ferromagnetic material, the internal magnetic induction is much larger than the external magnetic induction. One may also say that the magnetic induction lines are diluted in diamagnetic materials, concentrated in paramagnetic materials, and strongly concentrated in ferromagnetic materials.

In diamagnetic and paramagnetic materials, small applied fields give rise to an internal magnetic induction B_{int} that is directly proportional to the applied field strength

$$\vec{B}_{int} = \mu \vec{H} = \mu_r \mu_0 \vec{H}. \tag{8.17}$$

In order to find a relation between χ and μ, we consider a material placed in an external magnetic induction, $\vec{B}_0$, or an external magnetic field $\vec{H}$. The internal magnetic induction, $\vec{B}_{int}$, can then be written as

$$\vec{B}_{int} = \vec{B}_0 + \mu_0 \vec{M} = \mu_0(\vec{H} + \vec{M}), \tag{8.18}$$

provided demagnetization effects are neglected and the internal magnetic field $\vec{H}_{int}$ is approximated by the external magnetic field $\vec{H}$. For diamagnetic or paramagnetic materials, this approximation is justified and, after combining Eqs. (8.17) and (8.18), one finds

$$M = (\mu_r - 1)H = \chi H \tag{8.19}$$

or

$$\mu_r = \chi + 1, \tag{8.20}$$

where χ is the (dimensionless) volume susceptibility. For ferromagnetic materials, it is not justified to approximate H_{int} by the applied field H in Eq. (8.18). In ferromagnetic materials, strong demagnetizing fields are present below the Curie temperature with field strengths that are commonly much larger than the applied fields. Instead of Eq. (8.18), we therefore write

$$\vec{B}_{int} = \mu_0(\vec{M} + \vec{H}_d), \tag{8.21}$$

where $\vec{H}_d$ is the demagnetizing field, and where we have assumed zero external field. The existence of a demagnetizing field can best be understood by considering a bar magnet for which the magnetic induction B and the magnetic field inside and outside the magnet are as shown in Fig. 8.1. Inspection of this figure shows that the field lines and flux lines outside the bar magnet are the same, which is plausible since in free space we have $\vec{B} = \mu_0 \vec{H}$.

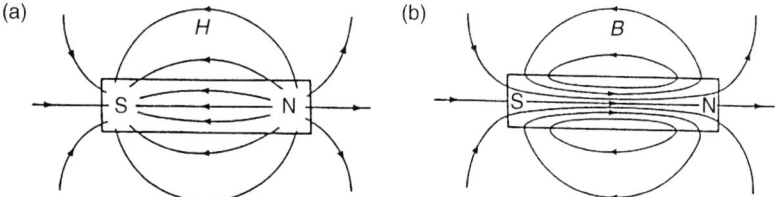

Figure 8.1. (a) Magnetic field H inside and outside a bar magnet, (b) magnetic induction B inside and outside a bar magnet.

On the basis of Maxwell's equations applied to a situation in which there are no electric currents, one has

$$\text{curl } \vec{H} = 0, \tag{8.22}$$

and for the flux density $\vec{B}$ one has

$$\text{div } \vec{B} = 0. \tag{8.23}$$

For a bar magnet of finite dimensions, one may therefore write

$$\int_{\text{space}} (\vec{H} \cdot \vec{B}) \, dV = 0, \tag{8.24}$$

where the integration is performed over the whole space. This integral may be written as the sum of the integral over the volume of the bar magnet and the integral over the rest of the (free) space

$$\int_{\text{magnet}} (\vec{H} \cdot \vec{B}) \, dV = -\int_{\text{rest}} (\vec{H} \cdot \vec{B}) \, dV = -\mu_0 \int_{\text{rest}} H^2 \, dV. \tag{8.25}$$

This result shows that the integral over the volume of the magnet (first term in Eq. 8.25) has to be negative, which is possible only if $\vec{H}$ and $\vec{B}$ inside the magnet have opposite direction. In other words, inside the magnet exists a magnetic field with a direction opposite to that of the magnetization and hence the name demagnetizing field.

The demagnetizing field $\vec{H}_d$ depends on the shape of the magnet. For a homogeneously magnetized ellipsoid it can be expressed as

$$\vec{H}_d = -N_d \vec{M}, \tag{8.26}$$

where the demagnetization factor N_d is dimensionless with values ranging between zero and one. This factor is a sensitive function of the geometry of the magnet. Examples of demagnetizing factors pertaining to shapes of simple geometry are listed in Table 8.1.

Using Eqs. (8.21) and (8.26), one has for the induction inside the magnet

$$\vec{B}_{\text{int}} = \mu_0 (\vec{M} - N_d \vec{M}) = \mu_0 \vec{M}(1 - N_d). \tag{8.27}$$

The units used for describing the magnetic properties of the various magnetic materials considered in the literature are far from being uniform. Throughout this book, the Standard International system of units (SI) has been used, that was adopted in 1960 by the Conférence

Table 8.1. Demagnetizing factors for some magnetic bodies of simple geometries, including cylinders with various length (l) to diameter (d) ratios

Geometry	l/d	N_d
Toroid		0
Long cylinder		0
Cylinder	20	0.006
Cylinder	10	0.017
Cylinder	5	0.040
Cylinder	1	0.27
Sphere		0.333

Générale des Poids et Mesures. These units form a coherent system and are based on seven basic units: meter, kilogram, second, ampere, kelvin, mole, and candela. The use of the SI units has been recommended by the International Union of Pure and Applied Physics (Cohen and Giacomo, 1987). In order to make easy contact with the enormous amount of magnetic data published in the scientific literature during the years, many scientists still use the older cgs–emu units even at present. For this reason, we have listed in Table 8.2 relationships and conversion factors between the SI and the older units.

The flux density B is not always a good measure to characterize a magnetic material since we have seen that it may include contributions from external magnetic sources. The intrinsic properties of a given material are therefore always characterized by the magnetization M (in A m^{-1}) or the magnetic polarization J (in T). The following relation exists between $\vec{M}$ and $\vec{J}$:

$$\vec{J} = \mu_0 \vec{M}. \tag{8.28}$$

The flux density $\vec{B}$, polarization $\vec{J}$, and field strength $\vec{H}$ are related by the equation

$$\vec{B} = \vec{J} + \mu_0 \vec{H}. \tag{8.29}$$

Ferromagnetic materials are characterized by the presence of hysteresis loops. Examples of such loops are shown in Fig. 8.2. In so-called soft-magnetic materials, the loops are very narrow; in hard-magnetic materials the loops can be extremely broad. We will return to these points in Chapters 12–14. Here, we will restrict ourselves to a comparison of different types of representations of hysteresis loops.

Plots of B versus H and J versus H for a given ferromagnetic material are compared in Fig. 8.2. Both quantities B and J become zero at sufficiently high negative fields, which defines the corresponding coercive fields, indicated by $_BH_c$ and $_JH_c$, respectively. The field $_JH_c$ where the magnetic polarization or the magnetization vanishes is often referred to as the intrinsic coercivity. Many authors plot the magnetization M measured versus the corresponding magnetic field strength by using the symbol B of the flux density for the latter. In these particular cases, the flux density B is considered to represent an external quantity not related to the material under investigation. It is obtained by applying Eq. (8.29) to an empty measuring coil ($J = 0$) and leads to the relation $B_0 = \mu_0 H$, where B_0 can now be given in units of tesla. The advantage of this procedure is that field strengths given

CHAPTER 8. SOME BASIC CONCEPTS AND UNITS

Table 8.2. Magnetic quantities and conversion from Gaussian to SI units. For conversion between the two unit systems, multiply the number for the Gaussian quantity by the conversion factor to obtain the number for the SI quantity

Quantity	Symbol	Gaussian system	Conversion factor	SI system
Magnetic flux	Φ	Mx, G cm^2	10^{-8}	Wb, V s
Magnetic flux density, magnetic induction	B	G	10^{-4}	T, Wb m^{-2}
Magnetic potential difference, magnetomotoric force	U, F	Gb (gilbert)	$10/4\pi$	A
Magnetic field strength	H	Oe	$1000/4\pi$	A m^{-1}
Volume magnetization	$4\pi M$	G	$1000/4\pi$	A m^{-1}
Volume magnetization	M	emu cm^{-3}, G	1000	A m^{-1}
Magnetic polarization	J	emu cm^{-3}, G	$4\pi \times 10^{-4}$	T, Wb m^{-2}
Mass magnetization	M	emu g^{-1}, G cm^3 g^{-1}	1	A m^2 kg^{-1}, J T^{-1} kg^{-1}
Magnetic moment	m	emu, erg G^{-1}	10^{-3}	A m^2, J T^{-1}
Magnetic dipole moment	j	emu, erg G^{-1}	$4\pi \times 10^{-10}$	Wb m, V s m
Volume susceptibility	χ, κ	Dimensionless, emu cm^{-3}	4π	Dimensionless
Mass susceptibility	χ, κ	emu g^{-1}, cm^3 g^{-1}	$4\pi \times 10^{-3}$	m^3 kg^{-1}
Molar susceptibility	χ, mol	emu mol^{-1}, cm^3 mol^{-1}	$4\pi \times 10^{-6}$	m^3 mol^{-1}
Permeability, $\mu = \mu_0\mu_r$	μ	$\mu^* = \mu_r$	$4\pi \times 10^{-7}$	H m^{-1}, V s A^{-1} m^{-1}
Relative permeability, μ/μ_0	μ_r	Dimensionless	1	Dimensionless
Energy density, energy product	w	erg cm^{-3}	10^{-1}	J m^{-3}
Demagnetization factor	N, D	Dimensionless	$1/4\pi$	Dimensionless

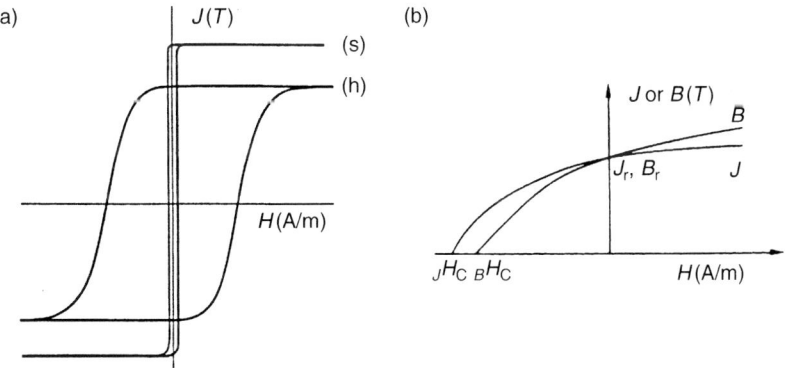

Figure 8.2. (a) Hysteresis loops observed in hard (h) and soft (s) ferromagnetic materials. (b) Comparison of B versus H plots and J versus H plots of the second quadrant of a hysteresis loop. In the SI system, B and J both are in T, and H in A m^{-1}. The remanence is indicated by B_r and J_r. The corresponding coercivities are represented by $_BH_c$ and $_JH_c$, respectively. Both loops were measured on long cylinders in order to exclude demagnetization effects (see Eq. 8.26 and Table 8.1).

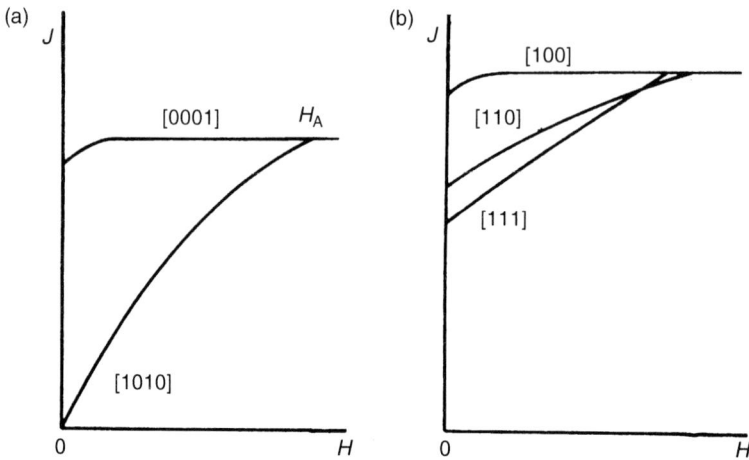

Figure 8.3. Schematic representation of magnetization curves measured in various crystallographic directions for (a) Co and (b) Fe.

in kOe in the older units can be compared easily with values of $\mu_0 H$ in tesla in modern SI units.

Magnetically hard materials are characterized by a strong preference of the magnetization for one of the crystallographic directions. In this direction, the full magnetization is already reached in comparatively low field strength, whereas much higher field strengths are needed in the other (hard) directions. Examples of measurements revealing this property are shown in Fig. 8.3. It will be clear that the difference in magnetic response between the various directions is more distinct in plots of J or M versus H than in plots of B versus H. In order to exclude demagnetization effects, such measurements are preferentially made on long cylinders of the material with the long axis cut in the various crystallographic directions considered.

Before leaving this chapter we will briefly mention the existing dilemma between the primacy of $\vec{B}$ and $\vec{H}$. For this reason we have summarized Maxwell's equations and several other related equations in Table 8.3. When inspecting Maxwell's equations one may notice two types of analogies between the electric and magnetic quantities: (i) The distinction according to the structure of the differential operators (curl and div) leads to the correspondences $\vec{E} \Leftrightarrow \vec{H}$ and $\vec{D} \Leftrightarrow \vec{B}$. In this picture, the primacy of H is based on fictive magnetic surface poles or *magnetic charges* traditionally used in magnetostatics by analogy with electrostatics; (ii) by contrast, the distinction by homogeneous and inhomogeneous equations leads to the correspondences $\vec{E} \Leftrightarrow \vec{B}$ and $\vec{D} \Leftrightarrow \vec{H}$, suggesting that it is more appropriate to consider the magnetic induction or magnetic flux density $\vec{B}$ as the field in matter.

In the codification of the field quantities of Faraday, Maxwell, and their contemporaries, the primary magnetic field is $\vec{H}$ taken as the *magnetizing force*, now called field strength. The term *magnetic induction* for $\vec{B}$ points to the status of a variable quantity that is determined by the sum of $\vec{H}$ and $\vec{M}$ and corresponds to the analogy mentioned under point (i) above. However, as stated in (ii) above, the field within magnetized matter is not $\vec{H}$ but $\vec{B}$, since

Table 8.3. Maxwell's and related equations. $\vec{E}$ electric field; $\vec{P}$ electric polarization; $\vec{H}$ magnetic field; $\vec{B}$ magnetic flux density; $\vec{M}$ magnetization; χ_m, χ_e magnetic or electric susceptibility; μ permeability; ε permittivity

$$\text{curl } \vec{E} = -\partial \vec{B}/\partial t$$
$$\text{curl } \vec{H} = \vec{j} + \partial \vec{D}/\partial t$$
$$\text{div } \vec{D} = \rho$$
$$\text{div } \vec{B} = 0$$
$$\vec{D} = \varepsilon \vec{E} = \varepsilon_0 \varepsilon_r \vec{E} = \varepsilon_0 \vec{E} + \vec{P}$$
$$\vec{B} = \mu \vec{H} = \mu_0 \mu_r \vec{H} = \mu_0(\vec{H} + \vec{M})$$
$$\varepsilon_r = 1 + \chi_e$$
$$\mu_r = 1 + \chi_m$$

$\vec{B}$ is produced by both the conduction and the Amperian (atomic) currents, while $\vec{H}$ has its source exclusively in external conduction currents. Hence, the analogy considered under point (ii) appears to be more general. It is clear that those advocating the primacy of $\vec{B}$ have an excellent viewpoint although historically they are at a disadvantage, because the magnetic pole approach has traditionally been used in magnetostatics by analogy with electrostatics. In fact, for the characterization of magnetic materials the traditional approach is usually preferred because the desired quantity is $M(H, T)$. The relevant field $\vec{H}$ in a magnetized sample is the internal field, $\vec{H}_{int}$, which originates not only from the external field $\vec{H}_{ext}$, but also from the stray- or self-field of the magnetized body. Thus, $\vec{H}_{ext}$ has to be corrected by the demagnetizing field $\vec{H}_d = -N_d \vec{M}$, where N_d is the demagnetizing factor and depends on the sample geometry, as shown in Table 8.1. A more detailed discussion regarding these matters can be found in articles written by Cohen and Giacomo (1987), Goldfarb (1992), and Hilscher (2001).

References

Cohen, E. R. and Giacomo, P. (1987) Symbols, units, nomenclature and fundamental constants in physics, *Physica A*, *146*, 1.

Goldfarb, R. B. (1992) Demagnetizing factors, in J. Evetts (Ed.) *Concise encyclopedia of magnetic and superconducting materials*, Oxford: Pergamon, p. 103–104.

Hilscher, G. (2001) in *Encyclopedia of materials: science and technology*, Amsterdam: Elsevier Science.

9

Measurement Techniques

9.1. THE SUSCEPTIBILITY BALANCE

Commonly, for measuring the magnetic susceptibility of samples having the shape of long prismatic or cylindrical rods, the Gouy method is applied. A schematic representation of a measuring device based on this method is shown in Fig. 9.1.1.

The sample is suspended using a long string and a small counterweight to prevent the sample from being pulled to one of the magnet poles. One end of the rod-shaped sample is located between the poles of a magnet where the field strength is comparatively high. By contrast, the field strength at the other end of the sample is small. The axial force exerted on the sample by the field is measured, for instance, by using an automatic balance.

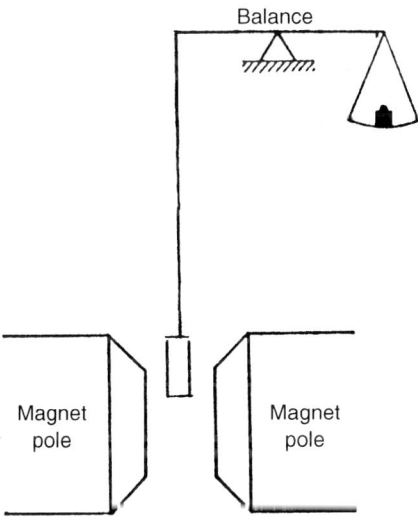

Figure 9.1.1. Schematic diagram of a susceptibility balance. After Zijlstra (1967).

If the long axis of the rod is along the z-axis, we are interested in the axial force dF_z exerted on a volume element dV of the sample. Using the relations

$$\vec{F} = -\text{grad } E \quad \text{and} \quad E = -\mu_0 \vec{\mu} \cdot \vec{H},$$

where E represents the magnetostatic energy of the sample in the magnetic field, we may write

$$dF_z = \frac{\partial \left(\mu_0 \vec{\mu}_V \cdot \vec{H}\right)}{\partial z}, \tag{9.1.1}$$

where μ_V is the magnetic moment of the volume element considered and H is the corresponding local field strength. If the sample is homogeneous, all volume elements of the sample have the same magnetic moment given by $\mu_V = \chi H$. For paramagnetic and diamagnetic samples, it has been shown in Chapters 3 and 6 that the susceptibility is field-independent. In that case, one may write

$$dF_z = \frac{1}{2}\mu_0 \chi \frac{\partial H^2}{\partial z} dV. \tag{9.1.2}$$

After integration along the length of the sample, one finds for the total axial force F_z

$$F_z = \int dF_z = \frac{1}{2}\mu_0 \chi a \int \frac{\partial H^2}{\partial z} dz = \frac{1}{2}\mu_0 \chi a \left(H_b^2 - H_t^2\right), \tag{9.1.3}$$

where a is the cross-sectional area of the rod-shaped sample perpendicular to the z-axis, H_b is the field strength at the bottom of the sample located between the magnet poles, and H_t is the field strength at the top of the sample.

It follows from Eq. (9.1.3) that the force is independent of the direction of H_t and H_b. If H_t is smaller than one tenth of H_b, its neglect leads to an error of at most 1%.

The Gouy method works satisfactorily if the susceptibility is isotropic and field-independent. The sample rod has to be macroscopically homogeneous and a constant cross-section is required. Often, the sample consists of a glass tube filled with powder. In this case, one has to prevent inhomogeneous compression by the field, which can be done by fixing the powder particles by means of glue.

In the Gouy method, one obtains the susceptibility by measuring the change of weight after the field in the magnet has been switched on. For practical purposes, it is sometimes convenient to calibrate the weight increase by means of a standard sample of well-known susceptibility.

9.2. THE FARADAY METHOD

In the Faraday method, the sample is again placed in an inhomogeneous magnetic field, the concomitant force being given by

$$F_z = \mu_0 \text{ grad}\left(\vec{\mu} \cdot \vec{H}\right) = \mu_0 \mu_z \frac{\partial H_z}{\partial z}, \tag{9.2.1}$$

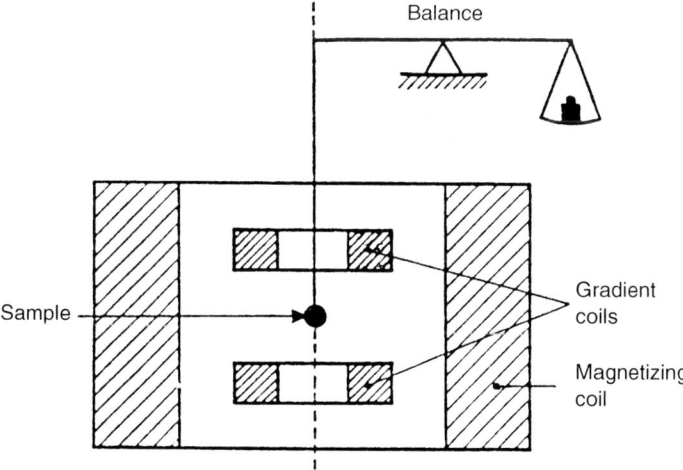

Figure 9.2.1. Schematic representation of an apparatus for measuring the magnetic moment of a small sample as a function of the applied field. After Zijlstra (1967).

where μ_z is the magnetic moment of a small sample located in an area where the field gradient $\partial H_z/\partial z$ has a constant value.

For various types of magnetically ordered materials, in particular, it is desirable to measure the magnetization as a function of the field strength H_0. In such cases, one may use an apparatus as schematically shown in Fig. 9.2.1, where a homogeneous magnetic field H_0 along the vertical or z-direction is generated by a solenoid. This field aligns the moment of the sample in the z-direction, and if the magnetization of the sample is field-dependent, the field applied will increase the magnitude of μ_z. For measuring the size of μ_z at each field strength H_0 by means of the force F_z (Eq. 9.2.1), one needs an auxiliary field gradient $\partial H_z/\partial z$, which can be generated by means of a set of gradient coils that are specially designed for producing a homogeneous gradient at the site of the (small) sample. The force F_z is measured, for instance, by an automatic balance. The force can be calibrated by measuring a standard sample of pure Ni, having a well-known magnetization.

9.3. THE VIBRATING-SAMPLE MAGNETOMETER

The vibrating-sample magnetometer (VSM) is based on Faraday's law which states that an emf will be generated in a coil when there is a change in flux linking the coil. Using Eqs. (8.9) and (8.10), we may write for a coil with n turns of cross-sectional area a:

$$V = -na\frac{dB}{dt}. \qquad (9.3.1)$$

If the coil is positioned in a constant magnetic field, one has

$$B = \mu_0 H.$$

When we bring a sample having a magnetization M into the coil, we have

$$B = \mu_0(H + M).$$

The corresponding flux change is

$$\Delta B = \mu_0 M. \quad (9.3.2)$$

Combining Eqs. (9.3.1) and (9.3.2) leads to

$$V dt = -na\mu_0 M. \quad (9.3.3)$$

This means that the output signal of the coil is proportional to the magnetization M but independent of the magnetic field in which the size of M is to be determined.

In the VSM, the sample is subjected to a sinusoidal motion (frequency υ) and the corresponding voltage is induced in suitably located stationary pickup coils. The electrical output signal of these latter coils has the same frequency υ. Its intensity is proportional to the magnetic moment of the sample, the vibration amplitude, and the frequency υ. A simplified schematic representation of the VSM is given in Fig. 9.3.1. The sample to be measured is centered in the region between the poles of a laboratory magnet, able to generate the measuring field H_0. A thin vertical sample rod connects the sample holder with a transducer assembly located above the magnet. The transducer converts a sinusoidal ac drive signal, provided by an oscillator/amplifier circuit, into a sinusoidal vertical vibration of the sample rod. The sample is thus subjected to a sinusoidal motion in the uniform magnetic field H_0.

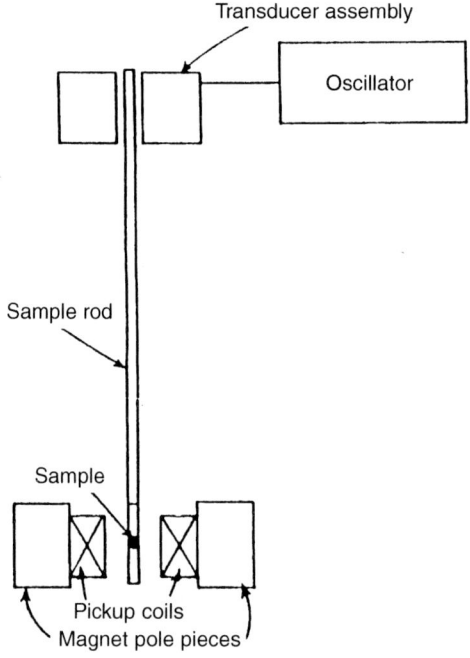

Figure 9.3.1. Schematic representation of a VSM.

Coils mounted on the poles of the magnet pick up the signal resulting from the motion of the sample. This ac signal at the vibration frequency υ is proportional to the magnitude of the moment of the sample. However, since it is also proportional to the vibration amplitude and frequency, the moment readings taken simply by measuring the amplitude of the signal are subject to errors due to variations in the amplitude and frequency of vibration. In order to avoid this difficulty, a nulling technique is frequently employed to obtain moment readings that are free of these sources of error. These techniques (not included in the diagram shown in the figure) make use of a vibrating capacitor for generating a reference signal that varies with moment, vibration amplitude, and vibration frequency in the same manner as the signal from the pickup coils. When these two signals are processed in an appropriate manner, it is possible to eliminate the effects of vibration amplitude and frequency shifts. In that case, one obtains readings that vary only with the moment of the sample.

9.4. THE SQUID MAGNETOMETER

The influence of magnetic flux on a Josephson junction may be employed for measuring magnetic fields or magnetizations. The basic element of a Superconducting Quantum Interference Device (SQUID) magnetometer is a ring of superconducting metal containing one or two weak links. The name quantum interference is derived from the fact that the critical current of an array of two Josephson junctions is periodic in field units of $h/2e$ due to interference effects of the electron-pair wave functions. A so-called dc SQUID is built with two Josephson junctions and a dc current is applied to this device. The effect of a radio frequency (RF) field on the critical current is used to detect quasi-static flux variations. The RF SQUID is a simple ring with only one Josephson junction. Variation of the flux in the ring results in a change of impedance. This change in impedance results in detuning of a weakly coupled resonator circuit driven by an RF current source. Therefore, when a magnetic flux is applied to the ring, an induced current flows around the superconducting ring. In turn, this current induces a variation of the RF voltage across the circuit. With a lock-in amplifier this variation is detected. A feedback arrangement is used to minimize the current flowing in the ring, the size of the feedback current being a measure of the applied magnetic flux. The detection by means of a SQUID is extremely sensitive. In commercial magnetometers the method is capable of measuring magnetic moments in the range $10^{-10}-10^3$ A m^2 with an accuracy of 1%. Custom-designed dc SQUIDs can have a few orders of magnitude higher sensitivities. For a detailed treatise on the operation principles and design considerations of dc and RF SQUID sensors, we refer to the book of van Duzer and Turner (1981) or the chapter by Clarke (1977).

References

Clarke, J. (1977) in B. B. Schwartz and S. Foner (Eds) *Superconductor applications: SQUIDS and machines*, New York: Plenum Press.
van Duzer, T. and Turner, C. W. (1981) *Principles of superconductive devices and circuits*, New York: Elsevier.
Zijlstra, H. (1967) *Experimental methods in magnetism*, Amsterdam: North Holland Publishing Company.

10

Caloric Effects in Magnetic Materials

10.1. THE SPECIFIC-HEAT ANOMALY

It has been shown in Chapter 4 that the magnetic susceptibility of ferromagnets, ferrimagnets, and antiferromagnets behaves anomalously when the temperature in the paramagnetic range approaches the magnetic-ordering temperature. In this section, it will be shown that the anomalies in magnetic behavior close to the ordering temperature are accompanied by anomalies in the specific heat.

Let us first consider the effect of an external field H on a magnetic material for which the magnetization is equal to zero before a magnetic field is applied. The work necessary to magnetize a unit volume of the material is given by

$$\partial W = \mu_0 H \, \partial M.$$

The total work required to magnetize a unit volume of the material is

$$W = \int \mu_0 H \, dM. \tag{10.1.1}$$

In analogy with Eq. (10.1.1), one finds that the spontaneous magnetization of a ferromagnetic material gives rise to an additional contribution to the internal energy per unit volume of the material

$$E_m = -\int \mu_0 H_m \, dM, \tag{10.1.2}$$

where $H_m = N_W M$ is the molecular field or Weiss field introduced in Chapter 4. After substituting $N_W M$ for H_m and performing the integration from 0 to M, one obtains

$$E_m = -\tfrac{1}{2} \mu_0 N_W M^2. \tag{10.1.3}$$

The additional specific heat due to the spontaneous magnetization is then given in the molecular field approach by

$$C_m = \frac{dE_m}{dT} = -\frac{1}{2} \mu_0 N_W \frac{dM^2}{dT}. \tag{10.1.4}$$

In writing down these equations, one has to realize that M depends on temperature and that dM^2/dT varies strongly with temperature. When inspecting Fig. 4.2.1c, one sees that $M = 0$ in a ferromagnet material above T_C whereas M is almost temperature-independent at temperatures much below T_C. However, M varies strongly just below T_C. In terms of Eq. (10.1.4), this means that C_m vanishes at very low temperatures and above T_C.

Just below T_C, the specific heat will be large. In fact, dM^2/dT shows a discontinuity at T_C. The size of this discontinuity can be calculated as follows. The molecular field constant N_W can be expressed in terms of T_C by rewriting Eq. (4.2.5) as:

$$N_W = \frac{3kT_C}{N\mu_0 g^2 J(J+1)\mu_B^2}. \tag{10.1.5}$$

In Section 4.2, it has already been shown that the reduced magnetization $M(T)/M(0)$ if plotted as a function of the reduced temperature has the same shape for all ferromagnetic materials characterized by the same quantum number J. By substituting $M(T)/M(0)$ of Eq. (4.2.11) into Eq. (10.1.5), one can calculate C_m exactly over the whole temperature range from $T/T_C = 1$ to $T/T_C = 0$ by means of a simple computational procedure.

If one is only interested in the magnitude of the specific-heat discontinuity at T_C, one may write down a series expansion for B_J of Eq. (4.2.1) and retain only the first two terms (Eq. 3.2.1). After some algebra, one finally finds for the magnitude of the discontinuity at T_C

$$\Delta C_m = \frac{5J(J+1)Nk}{J^2 + (J+1)^2}. \tag{10.1.6}$$

It is useful to keep in mind that for the simple case $J = \frac{1}{2}$ the specific heat jump at T_c equals $\Delta C_m = \frac{3}{2}Nk = \frac{3}{2}R$ for a mole of magnetic material. The temperature dependence of C_m for the case $J = \frac{1}{2}$ is shown in Fig. 10.1.1.

It is instructive to compare the molecular field results shown in Fig. 10.1.1 with the experimental results obtained for nickel, shown in Fig. 10.1.2. The upper curve in Fig. 10.1.2 is the total specific heat C_p. In order to compare this quantity with the molecular field prediction, one has to subtract the non-magnetic contributions due to lattice vibrations, thermal expansion, and the electronic specific heat. These non-magnetic contributions may

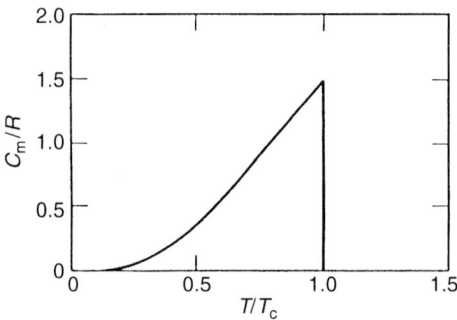

Figure 10.1.1. Magnetic contribution to the specific heat for a $J = \frac{1}{2}$ ferromagnet. After Morrish (1965).

SECTION 10.2. THE MAGNETOCALORIC EFFECT

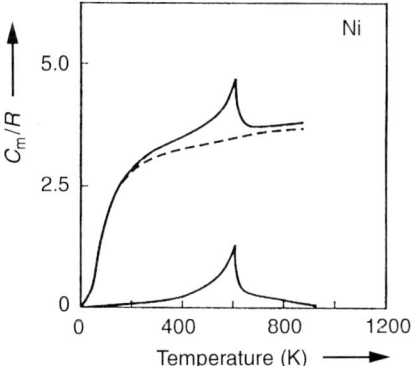

Figure 10.1.2. Temperature dependence of the specific heat of Ni (after Hofman et al. (1956)). The lower part of the figure represents the magnetic contribution obtained from the total specific heat after subtraction of the other contributions (broken line).

be estimated from measurements on nickel alloys that show no magnetic ordering. The total of these contributions has a temperature dependence as shown by the broken line in Fig. 10.1.2. After subtraction of this contribution, one finds the magnetic contribution shown in the lower part of the figure. Comparison with the molecular field result in Fig. 10.1.1 shows that the general behavior is the same, the main difference being substantial contributions also above T_C in the experimental curve. This behavior is commonly attributed to so-called short-range magnetic order. Above T_C, the long-range magnetic order that extends over many interatomic distances disappears. Some short-range order in terms of correlations between the directions of moments of nearest-neighbor atoms may persist, however, also at temperatures above the magnetic-ordering temperature.

10.2. THE MAGNETOCALORIC EFFECT

The magnetocaloric effect is based on the fact that at a fixed temperature the entropy of a system of magnetic moments can be lowered by the application of a magnetic field. The entropy is a measure of the disorder of a system, the larger its disorder, the higher its entropy. In the magnetic field, the moments will become partly aligned which means that the magnetic field lowers the entropy. The entropy also becomes lower if the temperature is lowered because the moments become more aligned.

Let us consider the isothermal magnetization of a paramagnetic material at a temperature T_1. The heat released by the spin system when it is magnetized is given by its change in entropy

$$\Delta Q = T_1 \Delta S. \quad (10.2.1)$$

If the magnetization measurement is performed under adiabatic conditions, the temperature of the magnetic material will increase. By the same token, if a magnetic material is adiabatically demagnetized, its temperature will decrease. The magnitude of the heat effects involved can be calculated as follows.

In the absence of the magnetic field, the $2J+1$ energy states of each of the participating magnetic moments are degenerate. For a system which contains N non-interacting magnetic moments, and therefore consists of $W = (2J+1)^N$ available states, we can easily calculate the entropy. According to Boltzmann's theory, the corresponding entropy is

$$S = k \ln W = Nk \ln(2J+1). \tag{10.2.2}$$

As has been discussed in Section 3.1 and illustrated for the case $J = \frac{9}{2}$ in Fig. 3.1.1, application of a magnetic field will lift the degeneracy of each of the N manifolds of $2J+1$ states. It follows from Fig. 3.1.1, and also from Eq. (3.1.1), that the energy separation between any two of the magnetically split $2J+1$ states equals $g\mu_0\mu_B H$. Let us suppose that the temperature T_1 at which the system is magnetized by means of the field H is so low that the thermal energy kT_1 is small compared to $g\mu_0\mu_B H$. In this case, only the lowest state ($m = -J$) will be occupied for each of the N spins. The corresponding entropy is now

$$S = Nk \ln 1 = 0. \tag{10.2.3}$$

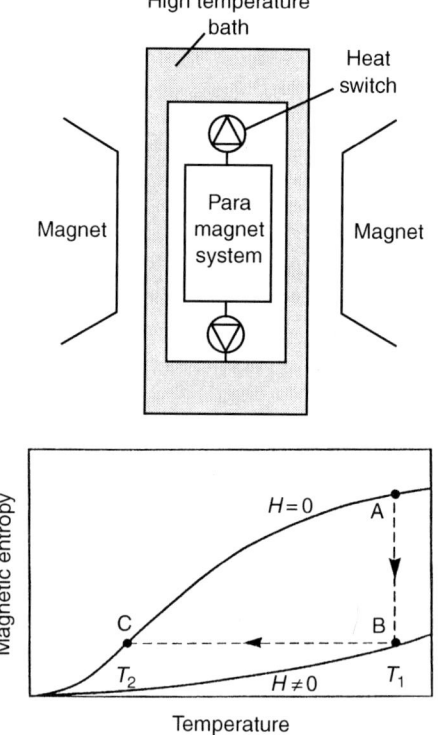

Figure 10.2.1. Schematic representation of a cooling system based on adiabatic demagnetization of a paramagnetic system (top part) and of trajectories associated with isothermal magnetization [AB] and adiabatic demagnetization [BC] (bottom part).

Consequently, when the field H is applied under isothermal conditions, the heat released by the spin system is

$$\Delta Q = NkT_1 \ln(2J + 1). \tag{10.2.4}$$

The same amount of heat will be absorbed by the system during the demagnetization process after the field has been switched off.

The consecutive steps carried out in the cooling process are illustrated by means of Fig. 10.2.1. The field H is applied at the temperature T_1 when the paramagnetic system is in good thermal contact with the high-temperature bath (path AB in the lower part of Fig. 10.2.1). For instance, the thermal contact is in on-position when the space between the paramagnetic system and the high-temperature bath (which may be liquid hydrogen or liquid helium) is filled with helium gas. Subsequently, the paramagnetic system is thermally isolated by pumping the helium gas away (heat switch in off-position). Then, the magnetic field is also switched off. In the lower part of Fig. 10.2.1, this corresponds to path BC. The temperature has now dropped to T_2. The process described above is employed for the production of very low temperatures in the microkelvin range (Little, 1964).

References

Hofmann, J. A., Paskin, A., Tauer, K. J., and Weiss, R. J. (1956) *J. Phys. Chem. Sol.*, 1, 45.
Little, W. A. (1964) *Progress in cryogenics*, 4, 101.
Morrish, A. H. (1965) *The physical principles of magnetism*, New York: John Wiley and Sons.

11

Magnetic Anisotropy

A magnetic material is said to possess magnetic anisotropy if its internal energy depends on the direction of its spontaneous magnetization with respect to the crystallographic axes. Phenomenologically, the anisotropy energy, E_{an}, in a material with uniaxial (hexagonal and tetragonal) symmetry may be described by a series expansion. For tetragonal symmetry, the lowest order terms are given by

$$E_{an}(\theta, \varphi) = K_1 \sin^2 \theta + K_2 \sin^4 \theta + K_3 \sin^4 \theta \cos 4\varphi, \tag{11.1}$$

where K_1, K_2, and K_3 are the anisotropy constants and where the direction of the spontaneous magnetization relative to the single uniaxial (c-axis) direction and the a axis is given by the polar angles θ and φ, respectively (see Fig. 11.1).

In most cases, it is sufficient to consider only the K_1 and K_2 terms. The preferred magnetization direction will be along the c-axis in hexagonal or tetragonal crystal structures if K_1 predominates and $K_1 > 0$. It will be perpendicular to the c-axis if $K_1 < 0$. If K_1 is not predominant, the preferred magnetization may point in other directions. In the following, we will take only K_1 and K_2 into consideration. If one has a situation in which

$$K_1 > 0 \quad \text{and} \quad K_1 + K_2 > 0, \tag{11.2}$$

then one finds that the lowest value of the anisotropy energy is reached for $\theta = 0$, whereas if

$$K_1 < 0 \quad \text{and} \quad 2K_2 > -K_1, \tag{11.3}$$

the lowest anisotropy energy is reached for a θ value given by

$$\sin^2 \theta = -\frac{K_1}{2K_2}. \tag{11.4}$$

A diagram showing the preferred moment directions for different K_1 and K_2 values in a hexagonal crystal is given in Fig. 11.2.

Experimental values of anisotropy fields, H_{an}, are commonly obtained by measuring magnetic polarization curves with the field applied parallel and perpendicular to the easy magnetization direction. The anisotropy field is then obtained as the intersection of the two magnetization curves mentioned. Illustrative examples of measurements of H_{an}

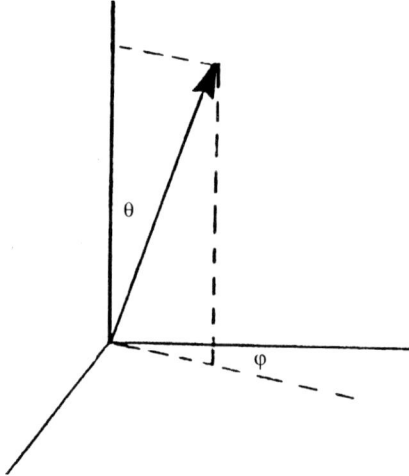

Figure 11.1. Moment direction relative to the c-axis defining the angles θ and φ.

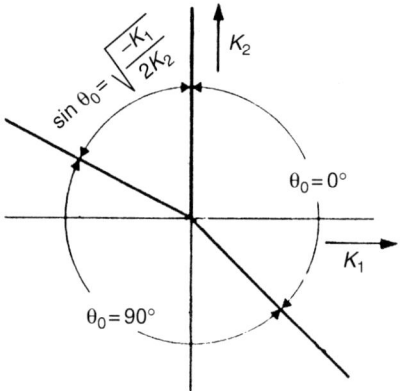

Figure 11.2. Diagram showing the preferred moment directions for different K_1 and K_2 values in a hexagonal crystal. After Smit and Wijn (1965).

obtained in this way for several permanent-magnet materials by Strnat (1988) are shown in Fig. 11.3.

Measurements of H_{an} can be helpful for obtaining an estimate of K_1. Suppose that the spontaneous polarization J_s of the material is held in equilibrium by a field H normal to the preferred direction, such that J_s is inclined at an angle θ and hence at an angle $90° - \theta$ with respect to H. The magnetic field H then exerts a torque $HJ_s \cos\theta$ that tends to increase θ. The torque tending to return J_s to the preferred direction is obtained by differentiating the expression for the anisotropy energy

$$\frac{dE_{an}}{d\theta} = 2K_1 \sin\theta \cos\theta + 4K_2 \sin^3\theta \cos\theta. \tag{11.5}$$

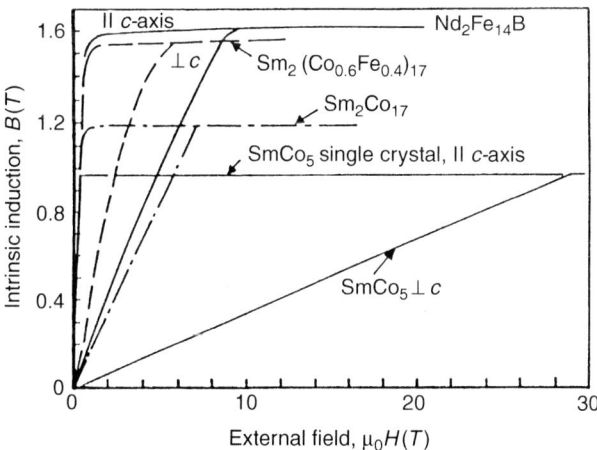

Figure 11.3. Easy-axis and hard-axis magnetization curves of several high-anisotropy compounds on which practical permanent magnets are based. After Strnat (1988).

Equating the two torques leads to the relation

$$H = \frac{2K_1 \sin\theta + 4K_3 \sin^3\theta}{J_s}. \tag{11.6}$$

The value of H that makes J_s parallel with the field is reached when $\sin\theta = 1$. The anisotropy field H_{an} is then given by

$$H_{an} = \frac{2K_1 + 4K_2}{J_s}. \tag{11.7}$$

In some materials, K_2 is negligible and measurements of H_{an} are sufficient for the determination of K_1.

A frequently used method to determine K_1 and K_2 has been developed by Sucksmith and Thompson (1954) and is based on the relation

$$\frac{2K_1}{J_s^2} + \frac{4K_2}{J_s^4}J^2 = \frac{H}{J}, \tag{11.8}$$

which holds for the magnetization curve of a single crystal obtained in comparatively small fields applied perpendicular to the easy direction. Under these circumstances, one may assume that the value of the saturation polarization J_s does not change with field strength and hence $\sin\theta = J/J_s$. Substitution into Eq. (11.6) then leads to Eq. (11.8). When H/J is plotted versus J^2, the anisotropy constant K_1 in Eq. (11.8) may be derived from the vertical intercept and the anisotropy constant K_2 from the slope of the straight line.

Substantial errors may arise from misalignment when this method is used for determining K_1 and K_2 on aligned powder samples. Misalignment leads to curvature of the magnetization, similar to what would be the effect of a larger value of K_2. Somewhat better in this respect is a method based on the Sucksmith–Thompson plot, as proposed by Ram and Gaunt (1983). In this modification, $H/\alpha(J - J_r)$ is plotted versus $\alpha^2(J - J_r)^2 \alpha^2(J - J_r)^2$,

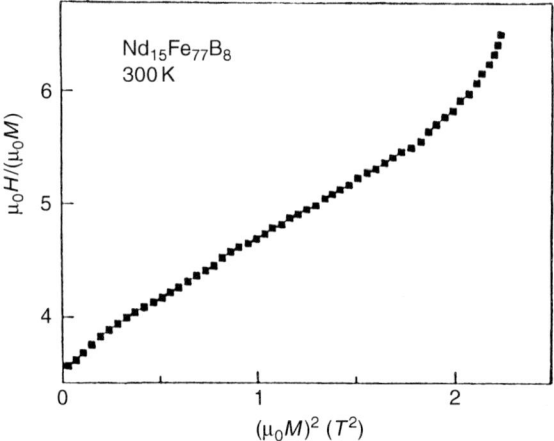

Figure 11.4. Sucksmith–Thompson plot for $Nd_{15}Fe_{77}B_8$ measured at room temperature. After Durst and Kronmüller (1986).

where J_r is the remanence in the hard direction and where the factor $\alpha = (J_s - J_r)/J_s$ has been introduced to simulate perfect magnetic alignment of the powder particles. An example of a modified Sucksmith–Thompson plot, obtained on a single crystal of $Nd_2Fe_{14}B$ by Durst and Kronmüller (1986), is shown in Fig. 11.4. The values of K_1 and K_2 derived from the intercept and slope in this plot are equal to 1.5 and 3.9 MJ m^{-2}, respectively.

The variation of the anisotropy energy with the direction of the magnetization in cubic materials is commonly expressed in terms of direction cosines. Let OA, OB, OC be the cube edges of a crystal and let the magnetization be in the direction of OP. Furthermore, $\alpha_1 = \cos POA$, $\alpha_2 = \cos POB$, and $\alpha_3 = \cos POB$. The anisotropy energy per unit volume of the material, if it is magnetized in the direction OP, is given by

$$E_{an} = K + K_4 \left(\alpha_1^2 \alpha_2^2 + \alpha_2^2 \alpha_3^2 + \alpha_3^2 \alpha_1^2 \right) + K_6 \left(\alpha_1^2 \alpha_2^2 \alpha_3^2 \right). \tag{11.9}$$

The constant K has been included for completeness, although it is rarely used. In many textbooks, the constants K_4 and K_6 are represented as K_1' and K_2'. Note that odd powers of α are absent in Eq. (11.9) because a change in sign of any of the αs should bring the magnetization vector into a direction that is equivalent to the original direction. Furthermore, the second-order terms can be left out of consideration since $\alpha_1^2 + \alpha_2^2 + \alpha_3^2 = 1$.

The anisotropy constants can most conveniently be determined by measuring the energy of magnetization $\int \vec{H} \, d\vec{J}$ along different crystal axes of a single crystal. These determinations include measurements of the $J(H)$ curve, starting from the demagnetized state up to magnetic saturation. Subsequently, the area between this curve and the J-axis is determined. Examples of such measurements were already displayed in Fig. 8.3.

The energies required for magnetizing cubic materials to saturation in the various crystallographic directions can be derived from Eq. (11.9). For the [100] direction, one

CHAPTER 11. MAGNETIC ANISOTROPY

obtains

$$\alpha_1 = 1 \quad \text{and} \quad \alpha_2 = \alpha_3 = 0. \tag{11.10}$$

In this case, $E_{100} = K$. In the face diagonal direction, [110], one obtains

$$\alpha_1 = \alpha_2 = \frac{1}{\sqrt{2}}, \quad \alpha_3 = 0. \tag{11.11}$$

Substitution of these values into Eq. (11.9) leads to $E_{110} = \frac{1}{4}K_4 + K$. In the same way, one finds for the [111] direction

$$\alpha_1 = \alpha_2 = \alpha_3 = \frac{1}{\sqrt{3}}. \tag{11.12}$$

After substitution into Eq. (11.9), one finds $E_{111} = \frac{1}{3}K_4 + \frac{1}{27}K_6 + K$. Combining these results leads to

$$K_4 = 4(E_{110} - E_{111}), \tag{11.13}$$

$$K_6 = 27(E_{111} - E_{100}) - 36(E_{110} - E_{100}). \tag{11.14}$$

These determinations of anisotropy constants have the advantage that possible errors due to strains are avoided, at least if these are isotropic and contribute equally to the energy of magnetization in all directions. It is also important that the energies are determined from curves between the remanence and the corresponding saturation value, rather than from initial magnetization curves because various domain processes not connected with crystalline anisotropy may contribute to the energy derived from the latter.

Other methods for determining the anisotropy constants make use of a torque magnetometer, by means of which it is possible to measure the torque, T, required to keep a crystal with its axes inclined at various known angles with respect to an applied magnetic field. In the ideal case, the measurements should be made with the sample cut in the shape of an oblate ellipsoid but a thin disc is usually satisfactory, provided a field well in excess of H_{an} can be applied. The disc is rotated around an axis perpendicular to both its plane and the applied field. It is most important that the sample have a circular shape and that it be mounted symmetrically about its center, because otherwise spurious torques will be introduced. It is difficult to interpret the results if the applied field does not saturate the sample (see the example given below). For this reason, the torque magnetometer is not frequently used for investigating permanent-magnet materials based on rare-earth elements that have very large anisotropies.

In cubic materials, the torque curves are expected to depend on the crystal plane of the sample. For a flat sample cut with its surface perpendicular to the [001] direction, one has for instance

$$\alpha_1 = \cos\theta, \quad \alpha_2 = \sin\theta \quad \text{and} \quad \alpha_3 = 0. \tag{11.15}$$

After substitution of these values into Eq. (11.9), one finds

$$E_{an} = K + K_4(\cos^2\theta \sin^2\theta) = K' + \frac{K_4}{8}(1 - \cos 4\theta). \tag{11.16}$$

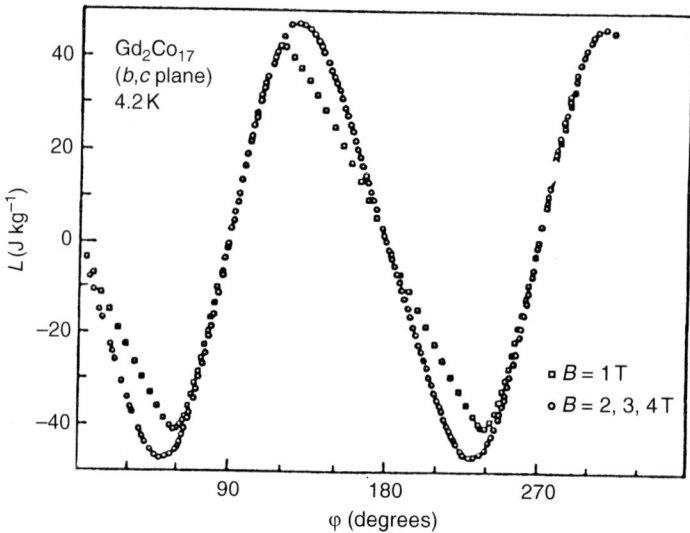

Figure 11.5. The torque measured in the b, c plane of Gd_2Co_{17} at 4.2 K for different applied magnetic fields. After Franse et al. (1989).

The torque can be obtained by differentiating this equation:

$$T = \tfrac{1}{2} K_4 \sin 4\theta. \tag{11.17}$$

The torque expressions for uniaxial symmetry are simpler and can be derived by differentiating Eq. (11.1). This leads to

$$T = K_1 \sin 2\theta + \tfrac{1}{2} K_2 \sin 4\theta. \tag{11.18}$$

Results obtained in this way by Franse et al. (1989) on a single crystal of the compound Gd_2Co_{17} are shown in Fig. 11.5. These measurements were made at 4.2 K. The easy magnetization direction in Gd_2Co_{17} is in a plane perpendicular to the hexagonal axis and the torque was measured in the b,c plane. It can be derived from the results shown that a magnetic field of 1 T is not strong enough to saturate the magnetization in the hard direction, that is, in the c-direction. After Fourier analysis of the curves and comparison with Eq. (11.18), the following values for the anisotropy constants are found: $K_1 = -29.5$ J kg^{-1}, $K_2 = -17.9$ J kg^{-1}, and $K_3 = 2.13$ J kg^{-1}.

More extensive descriptions of magnetic anisotropy and its determination can be found in the textbooks of Chikazumi (1966) and McCaig and Clegg (1987).

References

Chikazumi, S. (1966) *Physics of magnetism*, New York: John Wiley and Sons.
Durst, K. D. and Kronmüller, H. (1986) *J. Magn. Magn. Mater.*, 59, 86.
Franse, J. J. M., Sinnema, S., Verhoef, R., Radwanski, R. J., de Boer, F. R., and Menovsky, A. (1989) in I. V. Mitchell et al. (Eds) *CEAM Report*, London: Elsevier, p. 175.

McCaig, M. and Clegg, A. G. (1987) *Permanent magnets in theory and practice*, 2nd edn, London: Pentech Press.
Ram, V. S. and Gaunt, P. (1983) *J. Appl. Phys., 54*, 2872.
Smit, J. and Wijn, H. P. J. (1965) *Ferrites*, New York: Wiley.
Strnat, K. J. (1988) in E. P. Wohlfarth and K. H. J. Buschow (Eds) *Ferromagnetic materials*, Amsterdam: North Holland, *Vol. 4*, p. 131.
Sucksmith, W. and Thompson, J. E. (1954) *Proc. Roy. Soc. London, A225*, 362.

12

Permanent Magnets

12.1. INTRODUCTION

Permanent-magnetic materials are characterized by a field dependence of the magnetization showing a broad hysteresis loop and a concomitant high coercivity. The remanence B_r determines the flux density that remains after removal of the magnetizing field and hence is a measure of the strength of the magnet, whereas the coercivity $_BH_c$ is a measure of the resistance of the magnet against demagnetizing fields (see Fig. 8.2). The performance of a magnet is usually specified by its energy product, defined as the product of the flux density B and the corresponding opposing field H. If the hysteresis loop for a given magnet material is available, the energy product of a particular magnet body made of this material can be derived relatively easily. We illustrate this by means of Fig. 12.1.1, where we compare two different types of magnet materials (A and B). In the left panels of the figures, the second quadrants of the hysteresis loops of the two magnet materials are shown. In both cases, these loops have been measured on samples of the magnet materials having the form of long cylinders so that demagnetizing effects can be neglected ($N_d = 0$, see Table 8.1).

In the second quadrant, the direction of the external field is opposite to the flux density. Each point on the B–H curve can be taken to represent the working point of a magnet body subjected to its own demagnetizing field. Small demagnetizing fields and working points close to the B axis apply in general to elongated or rod-shaped (the length of the rod being large compared with its diameter) magnet bodies in their own demagnetizing field. By contrast, the working points of a magnet body with a flat or disk-like shape correspond to much larger demagnetizing fields and hence are located closer to the H axis. The energy products BH for low or high demagnetizing fields, that apply to the two mentioned types of magnet shapes, are relatively small as can be derived from the low values of the surface area of the corresponding BH rectangles. The energy products (horizontal scale) corresponding to all points of the $B(H)$ curve are plotted as a function of the flux density (vertical scale) in the right-hand parts of the figure. The largest possible value of the energy product for each magnetic material is indicated by $(BH)_{max}$. The corresponding working points are indicated on the $B(H)$ curves of both magnet materials as a filled and an open circle.

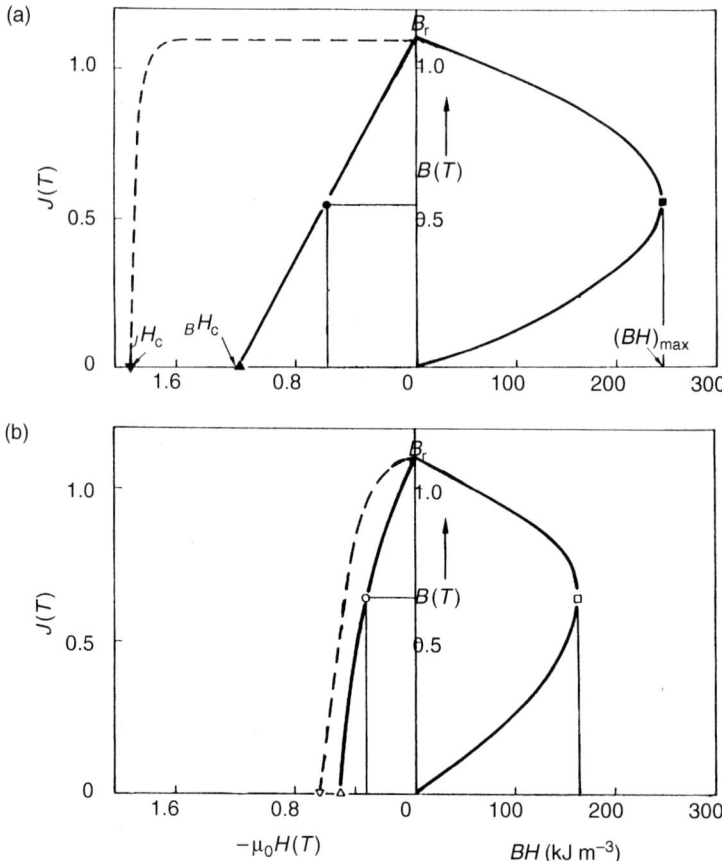

Figure 12.1.1. Comparison of the hard-magnetic properties of two hard-magnetic materials. Left-hand side: Flux density B (full lines) and magnetic polarization J (broken lines) as a function of the demagnetizing field strength H. Right-hand side: Product BH (horizontal axis) as a function of B (vertical axis) for both materials. The working point corresponding to $(BH)_{max}$ is indicated on the $B(H)$ curve for the materials A and B by a filled and an open circle, respectively. The unit of the energy product is kJ m^{-3}, if B is given in T and H in kA m^{-1}.

12.2. SUITABILITY CRITERIA

The maximum energy product is one of the most generally used criteria for characterizing the performance of a given permanent-magnet material. The magnitude of this product can be shown to be equal to twice the potential energy of the magnetic field outside the magnet divided by the volume of the magnet.

The maximum energy product is not the only criterion that can be used to specify the quality of a given permanent-magnet material. Of importance in many static applications is, for instance, the magnitude of the intrinsic coercivity $_JH_c$. This is illustrated in Fig. 12.1.1, which compares the $J(H)$ and $B(H)$ curves of two different magnet materials that have different hysteresis loops but the same remanence B_r. It follows from Eq. (8.29) that $_JH_c$

and $_BH_c$ will not be much different from each other when the former value is smaller than the remanence of the permanent-magnet material, as for the material B in Fig. 12.1.1.

In permanent-magnet materials based on rare-earth compounds, the intrinsic coercivity $_JH_c$ can become much larger than the remanence. This situation is illustrated by the example of material A shown in Fig. 12.1.1. In this case, the value of the intrinsic coercivity considerably exceeds the field corresponding to the $(BH)_{max}$ point and also exceeds $_BH_c$, the field where the magnetic flux vanishes. In the following, it will be assumed that both magnets form part of a magnetic circuit and that they have a shape corresponding to their $(BH)_{max}$ point. When incorporated into a magnetic circuit, in which external magnetic fields are present, the magnet material B in Fig. 12.1.1 is able to resist only a relatively small demagnetizing field. For instance, magnetizing fields higher than twice the field corresponding to the $(BH)_{max}$ point will completely demagnetize the magnet body and hence make it useless. In contrast, the magnet material A in Fig. 12.1.1 is able to resist demagnetizing fields more than three times higher than the field at its $(BH)_{max}$ point. It may be seen from the figure that this behavior originates from the independence of the magnetic polarization J (broken line) on opposite external and/or internal fields up to a value close to $_JH_c$.

High values of $_JH_c$ can generally be obtained in magnet materials that have a high intrinsic magnetocrystalline anisotropy, as in rare-earth compounds. In materials where the hard-magnetic properties originate from shape anisotropy (Alnico-type materials, as will be discussed in more detail in Section 12.8), it is not possible to generate large coercivities. The $B(H)$ curves of representative rare-earth-based magnets are compared in Fig. 12.2.1 with the $B(H)$ curve of Ticonal XX (Alnico type) and with the $B(H)$ curves of some other common types of magnet materials. It is the presence of large coercivities in particular that makes the rare-earth-based magnets suitable for applications in which flat magnet shapes are required.

It follows from the foregoing that the $(BH)_{max}$ value itself is not always a sufficient criterion for the suitability of a given permanent-magnet material to be applicable in electric motors. More relevant to this case is the extent to which reverse fields can be applied that leave the magnetic properties of the magnet body unchanged after removal of these fields. The recoil line and the recoil energy are suitability criteria commonly used to characterize permanent-magnet materials for use in permanent-magnet devices in which substantial changes of the demagnetizing field occur in the air gap. For defining these quantities, one may consider a magnet body characterized by a $B(H)$ loop like the one shown in Fig. 12.2.2 ($\mu_0{_BH_c}$ smaller than B_r). After application of a demagnetizing field up to a value corresponding to point a, the material will generally not return along the line connecting a and B_r but along the line abc. This so-called recoil line has a slope similar to that of $B(H)$ in the first quadrant of the loop at B_r, that is, $\Delta B/\Delta H = \mu_0$. The hatched area in the figure (b is midway in between a and c) is commonly referred to as the recoil energy. This energy generally depends on the location of a, meaning that there is a maximum attainable value for each material. A relatively high value of the maximum recoil product is reached in magnet materials in which the high coercivity originates from a large magnetocrystalline anisotropy and where the recoil line coincides with the $B(H)$ curve over an extended field range. In magnets based on shape anisotropy, the maximum recoil energy is only a small fraction of $(BH)_{max}$.

Magnetic devices in which cyclic operations are involved and where reversibility plays a prominent role require quite a different criterion for the suitability of magnet materials. The

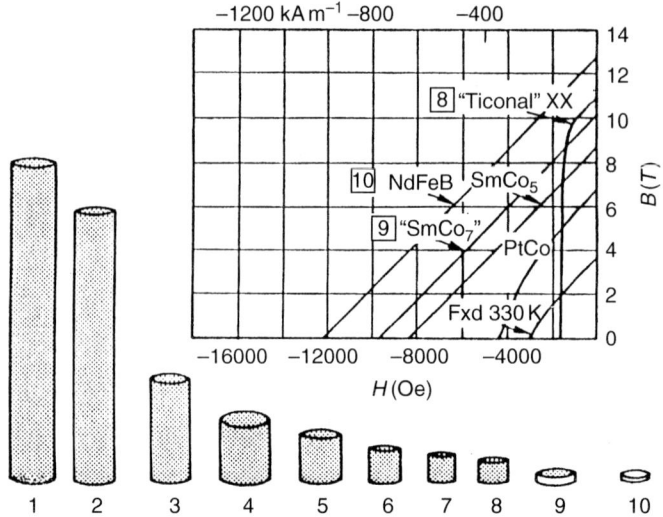

Figure 12.2.1. Schematic representation of magnet bodies made of different types of starting materials. The maximum energy products $(BH)_{max}$ increase steadily from left to right. The corresponding variation of B and H is reflected in the variation of the cross-sectional areas (ϕ) and in the variation of the lengths L of the magnets, respectively. The products $B\phi$ and HL are kept constant throughout the series, meaning that all these magnet bodies have the same magnetic flux and the ability to produce the same magnetic performance when incorporated in a magnetic circuit. The numbers refer to the following materials: 1. C-steel; 2. W-steel; 3. Co-steel; 4. Fe-Ni-Al-alloy; 5. "Ticonal II"; 6. "Ticonal G"; 7. "Ticonal GG"; 8. "Ticonal XX"; 9. SmCo$_7$; 10. Nd$_2$Fe$_{14}$B. B versus H plots of several magnet materials are shown in the top part of the figure. From Buschow (1988).

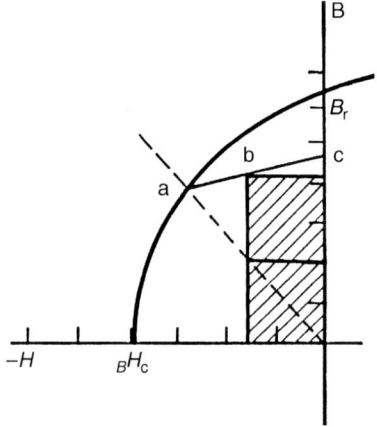

Figure 12.2.2. Recoil line abc and recoil energy (hatched area) for a given magnet material. From Buschow (1994).

relevant parameter here is the maximum amount of mechanical work that can be obtained in a reversible way from a well-designed configuration with a given magnet and a magnetizable object. It is well known that this maximum mechanical work (available per unit volume during a change in configuration) is equal to $J_s^2/2\mu_0$ in the case of an ideal magnet in which the complete (linear) hysteresis branch in the second quadrant is traversed reversibly.

There are also applications of permanent-magnet materials in which temporary or even cyclic excursions to elevated temperatures are required. In such cases, the suitability of a given magnet material will depend to some extent on the temperature dependence of its remanence and on the temperature dependence of its coercivity in the temperature range of interest. For many industrial applications, it is required to have stable coercivities and magnetizations up to at least 150°C. If both quantities decrease significantly with increasing temperature, one will be faced with a corresponding loss in magnet performance upon increasing the temperature. In the most favorable cases, these losses in magnet performance are only temporary and the original values of remanence and coercivity are recovered after returning to room temperature. Unfortunately, for some types of materials the loss in performance is irreversible. Reversible temperature coefficients of coercivity and remanence can usually be dealt with by designing a machine according to a given specification in a manner that the magnets are sized to be sufficiently strong at the highest temperature when they are most prone to demagnetization effects.

The corrosion resistance, the chemical and mechanical stability, the ease of mechanical processing, the weight per unit of energy product, and the electrical resistance are suitability criteria of a different kind that also have to be considered. Furthermore, one has to bear in mind that it is always necessary to magnetize magnets at some point in the manufacturing cycle. In favorable cases, this can conveniently be done with the magnets *in situ* in a partially or fully assembled machine, as with Alnico- and ferrite-type magnets. The production of machines in which premagnetized magnets are used may present severe problems. One of these is the attraction of magnetic dust during surface grinding. For this reason, it is sometimes desirable to employ magnets having coercivities that are sufficiently high for the purpose, but that are not so high as to make *in situ* magnetizing of the assembled magnet impossible. This means that the applicability of a magnetic material may require a lower as well as a higher limit for the coercivity. For more details, the reader is referred to the survey published by McCaig and Clegg (1987).

12.3. DOMAINS AND DOMAIN WALLS

It was mentioned already that not only a large maximum energy product $(BH)_{max}$, but also a high intrinsic coercivity $_JH_c$ is needed in some applications. Moreover, the maximum energy product itself depends on the coercivity and, if $_BH_c$ falls appreciably below the value J_s/μ_0, it may become lower than the theoretical limit

$$(BH)_{max} = \tfrac{1}{4} B_r {}_BH_c = \frac{J_s^2}{4\mu_0}. \tag{12.3.1}$$

For this reason, it is desirable to look somewhat more closely at the mechanisms that govern the magnitude of the coercivity in permanent-magnet materials.

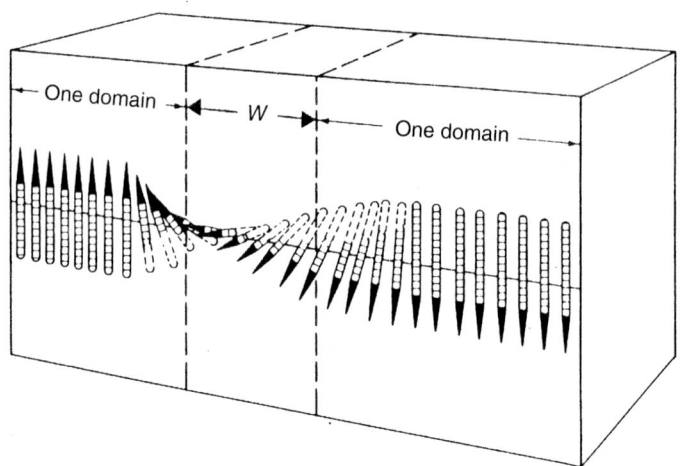

Figure 12.3.1. Schematic representation of the change in moment orientation within a 180° Bloch wall.

Let us consider the case that a steadily increasing magnetic field is applied in a direction opposite to the direction in which a perfect single crystal of a magnetic compound of hexagonal or tetragonal symmetry is magnetized (i.e., opposite to the easy magnetization direction). One would expect all atomic moments to reverse their direction by a process of uniform rotation only when the applied field becomes equal in size to the anisotropy field $H_{an} = 2K_1/M_s$ (see Chapter 11). However, in reality, such high coercivities are seldom encountered. Most permanent-magnet materials show the magnetization reversal already at field strengths that are only a small fraction (10–15%) of the value of $H_{an} = 2K_1/M_s$. The reason for this comparatively easy magnetization reversal is the existence of magnetic-domain structures. Magnetic particles of sufficiently large size will generally not be uniformly magnetized but rather be composed of magnetic domains that are mutually separated by domain walls or Bloch walls. A schematic representation of such a wall is given in Fig. 12.3.1. The magnetizations in adjacent domains point into opposite directions in order to reduce the magnetostatic energy. The magnetization in the wall between two domains gradually changes from the one preferred magnetization direction to the other. The thickness of the wall is determined by the relative strengths of the anisotropy energy and the exchange energy. The former tends to reduce the wall thickness, the latter tends to increase it. This may be seen from the argument given below.

According to Eq. (4.1.2), one may obtain the exchange energy between neighboring spins from the formula

$$H_{exch} = -2J_{exch}S^2 \sum_{i,j} \cos \varphi_{ij}, \qquad (12.3.2)$$

where φ_{ij} is the angle between the directions of the spin-angular-momentum vectors of atom i and its neighbors j. Generally, the widths of domain walls involve many lattice spacings, sometimes more than a hundred. For this reason, the angle φ_{ij} between two neighboring spins in the wall is very small, so that one may use the approximation $\cos \varphi_{ij} = 1 - \varphi_{ij}^2/2$. The variable part of the exchange energy for a row of atoms across the wall can then be

SECTION 12.3. DOMAINS AND DOMAIN WALLS

written as

$$H_{\text{exch}} = J_{\text{exch}} S^2 \sum_{i,j} \varphi_{ij}^2. \quad (12.3.3)$$

For two atoms with an angle φ between their spin moments, the variable part of the exchange energy is $E_{\text{exch}} = J_{\text{exch}} S^2 \varphi^2$ and would be equal to $E_{\text{abrupt}} = J_{\text{exch}} S^2 \pi^2$ if the magnetization would change abruptly over π radians. Let us consider the case that the directional change would be realized more gradually, and would involve N equal steps with equal angles π/N between neighboring spins. The total energy would then be only $E_{\text{gradual}} = J_{\text{exch}} S^2 (\pi/N)^2 N = J_{\text{exch}} S^2 \pi^2 / N$, which is a much smaller value. This result shows that the exchange energy favors a large wall width. However, the width is limited by the presence of the magnetocrystalline anisotropy energy, favoring collinear spin moments, being oriented in one of the two opposing easy directions. The actual width of the wall is determined by a competition between both energies.

A crude estimate of the energy and width of a domain wall can easily be obtained if we neglect the demagnetizing energy. Consider a 180° wall of width W in a simple cubic material extending along a given [100] direction in which the moment direction gradually changes from the positive to the negative [001] direction. Let us further assume that any deviation θ from the [001] direction involves an anisotropy energy given by $E = K \sin^2 \theta$. Within the wall, the moment directions largely deviate from the easy direction and the total anisotropy energy involved is roughly proportional to the wall width. If a is the lattice constant and if the wall extends over N lattice spacings, one obtains a rough estimate of the total anisotropy energy as $F_{\text{an}} = KW = KNa$, where $W = Na$ is the width of the wall. The increase in exchange energy for one row of atoms in the wall is $E = J_{\text{exch}} S^2 \pi^2 / N$. For a simple cubic lattice, the number of rows per unit area of wall is $1/a^2$. The exchange energy per unit area of wall is therefore $J_{\text{exch}} S^2 \pi^2 / Na^2$ or $J_{\text{exch}} S^2 \pi^2 / Wa$. The total energy per unit area associated with the wall is then

$$E_{\text{wall}} = \frac{J_{\text{exch}} S^2 \pi^2}{Wa} + K_1 W. \quad (12.3.4)$$

A minimum with respect to W is obtained when

$$\frac{\partial E_{\text{wall}}}{\partial W} = 0 = -\frac{J_{\text{exch}} S^2 \pi^2}{W^2 a} + K_1. \quad (12.3.5)$$

This leads to the following expression for the wall width W:

$$W = \sqrt{\frac{J_{\text{exch}} S^2 \pi^2}{K_1 a}} = \pi \sqrt{\frac{A}{K_1}}, \quad (12.3.6)$$

where A is the average exchange energy. Substitution of Eq. (12.3.6) into (12.3.4) leads to the following expression for the wall energy per unit area of a wall with width W:

$$E_{\text{wall}} = 2\pi \sqrt{\frac{J_{\text{exch}} S^2 \pi^2}{a}} = 2\pi \sqrt{K_1 A}. \quad (12.3.7)$$

In iron metal, one has $K \approx 5 \cdot 10^{-2}$ MJ m^{-3} and $a = 0.3$ nm. The value of J_{exch} may be calculated by means of Eq. (4.4.14), using $T_C = 1045$ K, $Z = 8$, and $S = 1.1$. This leads to $J_{\text{exch}} = 3 \cdot 10^{-21}$ J. By means of Eq. (12.3.6), one now finds

$$W \approx 50 \text{ nm},$$

which is about 200 lattice spacings, and

$$E_{\text{wall}} \approx 5 \cdot 10^{-3} \text{ J m}^{-2}.$$

This may be compared with the situation in a strongly anisotropic material like the tetragonal compound Nd$_2$Fe$_{14}$B for which $K_1 \approx 5 \cdot$ MJ m^{-3} and where the wall width is one order of magnitude smaller than in Fe metal.

12.4. COERCIVITY MECHANISMS

Already in 1948, Stoner and Wohlfarth showed that for a magnetization-reversal process proceeding by means of uniform rotation of the magnetic moments in spheroid particles, in which the major axis coincides with the easy direction of the magnetization, the coercivity is given by

$$H_c = \frac{2K_1}{M_s} - (N_\| - N_\perp) M_s, \qquad (12.4.1)$$

where $N_\|$ and $N_\perp$ are the demagnetizing factors corresponding to the two extreme directions of the spheroid particles. The first term is the normal anisotropy field that determines the easy magnetization direction when there is only magnetocrystalline anisotropy. The second term takes account of the fact that, even in the absence of magnetocrystalline anisotropy, the moments would align in the direction of the lowest demagnetizing factor. As already mentioned, the coercivity as expressed in Eq. (12.4.1) is based on a magnetization-reversal mechanism in which all moments retain their parallel arrangement during magnetization reversal (uniform rotation).

In practice, the coercivities obtained for most hard-magnetic materials are substantially lower, often by more than a factor of 10. This behavior is illustrated in Fig. 12.4.1, where deviations from the corresponding values of the nucleation field H_n, to be defined shortly, are shown, the latter representing the values of the first term of Eq. (12.4.1).

The reason for this is that there exists another magnetization-reversal mechanism that can proceed via considerably lower energy expenditure. The latter mechanism is based on nucleation of Bloch walls and growth of reversed domains. If, somewhere in a large single crystal, a tiny region with a less perfect magnetic-moment arrangement is present, it can serve to generate a Bloch wall. The Bloch wall will subsequently spread into the crystal and move across the whole crystal until magnetization reversal has been established over the whole crystal. Note that the energy required for this process is only equal to the wall energy taken over the whole surface of the wall and hence will involve only a very small volume compared to the total volume of the crystal. For the uniform-rotation process, the anisotropy energy taken over the whole volume of the crystal would be required.

Bloch walls and reversed domains can be generated near all types of defect regions where the local values of the exchange field and anisotropy field have become sufficiently

SECTION 12.4.   COERCIVITY MECHANISMS

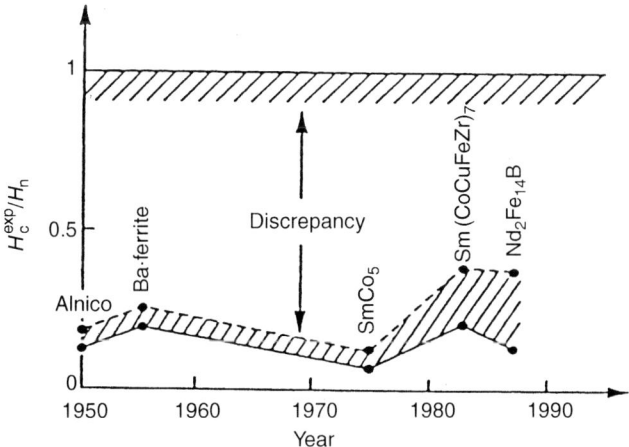

Figure 12.4.1. The ratio H_c/H_n of permanent magnets developed during the last four decades (dotted line: laboratory magnets, solid line: technical magnets). The values of H_n are equal to $2K_1/M_s$. After Kronmüller et al. (1988).

reduced with respect to the values in the bulk of the material to make a local magnetization reversal possible. This nucleation of Bloch walls at defects may take place spontaneously or under the influence of an externally applied negative magnetic field. The field required for Bloch-wall nucleation, commonly referred to as the nucleation field H_n, is often used to describe the concomitant coercivity H_c. Non-uniform processes in which magnetization reversal takes place by wall nucleation and propagation dominate in materials with high magnetocrystalline anisotropy. By analogy with Eq. (12.4.1), an empirical relation of the type

$$H_c = \alpha \frac{2K_1}{M_s} - N_{\text{eff}} M_s, \tag{12.4.2}$$

is often used to describe the nucleation field H_n and the concomitant coercivity H_c. The quantities α and N_{eff} are microstructural parameters that determine the relative importance of the magnetocrystalline anisotropy and the local demagnetizing field, respectively.

In the so-called nucleation-type magnet, the motion of the walls within the grains is comparatively easy. For obtaining high coercivities, the wall motion must be impeded by grain boundaries, since otherwise a single nucleated wall would lead to magnetization reversal of the entire magnet. The possibility of wall pinning at grain boundaries is therefore considered to be a prerequisite for nucleation-type magnets. Nucleation-type magnets may be characterized by the following properties: The low-field susceptibility, being a measure of the reversible displacement of walls, is very large. Magnetic saturation is already reached in comparatively low fields that are not much larger than the demagnetizing fields H_d. For obtaining the maximum coercivity, a positive saturation field ($H_s^{\max}$) of the order of the coercive field H_c is required. This necessity finds its origin in the possible persistence of residual domains of opposite magnetization up to $H_s^{\max}$. In fields larger than $H_s^{\max}$, all the walls will have been removed from the sample, except those walls that cannot be unpinned

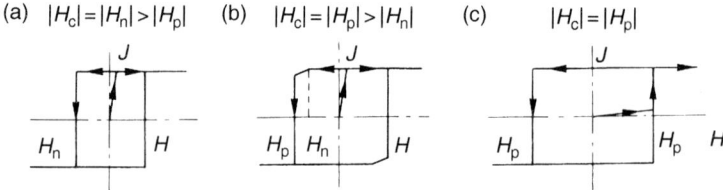

Figure 12.4.2. Schematic representation of the initial magnetization curve and hysteresis loop (a) for a material in which the nucleation mechanism is operative, (b) for a material in which the coercivity is determined by inhomogeneous nucleation and pinning of walls at grain boundaries, and (c) for a material in which the coercivity is controlled by wall pinning.

by any applied field. Generally, the value of the coercivity H_c cannot be further increased by application of positive fields larger than H_s^{max}.

After the application of H_s^{max}, nucleation of reversed domains can occur only in a negative field H at least equal in size to the nucleation field. Provided $|H_n|$ is larger than the propagation field $|H_p|$ associated with a possible wall pinning at the grain boundaries, complete magnetization occurs only if $|H| \geq |H_n|$, meaning that in this case the coercivity is equal to $|H_n|$ (see Fig. 12.4.2a).

A slightly different mechanism is also possible. Nucleation may take place at magnetic inhomogeneities at the grain boundaries where the propagation field associated with the pinning of walls at these inhomogeneities is larger than the nucleation field $|H_p| > |H_n|$. For intermediate field strengths $|H_p| > |H| > |H_n|$, nucleated domains may exist then but the domain walls will remain pinned at the grain boundary as long as $|H| < |H_p|$. This mechanism is commonly referred to as inhomogeneous pinning-controlled coercivity. It is difficult to distinguish it from the pure nucleation mechanism owing to the fact that the magnetization remains very close to saturation in fields $|H_p| > |H| > |H_n|$, because the volume of the domain nucleated (that has reversed magnetization) is negligibly small compared to the volume of the total grain. In Fig. 12.4.2b, the corresponding decrease in magnetization has been strongly exaggerated for clarity.

The situation is completely different in pinning-type magnets. Here the Bloch walls cannot travel freely throughout the whole grain because of magnetic inhomogeneities present in the grains that act as pinning centers for wall motion. Apart from the change in magnetization associated with some wall bending, this pinning will prevent further magnetization reversal. Wall displacement (other than bending) can occur only when the force exerted on the wall becomes sufficiently strong. This is the case when the strength of the external field exceeds the pinning field strength H_p that then determines the coercivity. A schematic representation of the hysteresis loop associated with such a situation is shown in Fig. 12.4.2c.

The presence of homogeneously distributed pinning centers inside the grain has important consequences for the low-field behavior. As illustrated in Fig. 12.4.2c, the low-field susceptibility is very weak. Saturation requires a field H_p sufficiently high to allow the walls to surmount the potential barriers associated with the pinning centers. The corresponding magnetization process is irreversible and it dominates any other reversible processes that may be present. The magnetization reversal occurring in a sufficiently high negative field ($|H_c|$) is subject to the same mechanism that takes place during the initial magnetization.

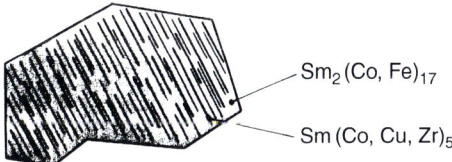

Figure 12.4.3. Schematic representation of the microstructure of a magnet material commonly referred to as "SmCo$_7$".

Consequently, the coercive field H_c is equal to the propagation field H_p that had shown up as a jump in the curve of initial magnetization. More details regarding the coercivity mechanisms described above can be found in reviews published by Zijlstra (1982), Givord et al. (1990), and Kronmüller (1991).

Permanent-magnet materials like "SmCo$_7$" are pinning controlled. At high temperatures, the alloy consists of one single phase. Heat treatment of the material at lower temperatures leads to the occurrence of a finely dispersed precipitate that is able to pin the Bloch walls and to cause high coercivities. A schematic representation of the microstructure of such a magnet material is shown in Fig. 12.4.3. In the permanent-magnet materials Nd$_2$Fe$_{14}$B and SmCo$_5$, the coercivity is nucleation controlled.

A survey of various magnet materials is given in Table 12.5.1. Extremely high coercivities are attained in all materials based on rare-earth elements. The reason for this is their high magnetocrystalline anisotropy discussed in Section 5.6, which leads to high coercivities in nucleation as well as in pinning controlled permanent magnets.

12.5. MAGNETIC ANISOTROPY AND EXCHANGE COUPLING IN PERMANENT-MAGNET MATERIALS BASED ON RARE-EARTH COMPOUNDS

The anisotropy in modern rare-earth-based magnet materials mentioned in the previous section derives primarily from the sublattice anisotropy of the rare-earth component R. The anisotropy of the 3d component is much weaker and, in some cases, has even the wrong sign, that is, it gives rise to an easy-plane magnetization. Generally, one may say that the rare-earth component in binary and ternary R-3d compounds is responsible for the magnetic anisotropy whereas the 3d component provides a sufficiently high magnetization and Curie temperature.

It follows from the results given in the previous section that high coercivities can be reached in materials in which the nucleation fields for domain walls H_n are high or in which the propagation fields H_p associated with domain-wall pinning are sufficiently high. It can be shown that both fields are the higher the stronger the magnetocrystalline anisotropy.

A discussion of the crystal field and the concomitant crystal-field-induced anisotropy has already been given in Chapter 5. We will now go a little further and show how the crystal-field parameters A_n^m that reflect the strength and symmetry of the crystal field are related to the macroscopic anisotropy constants K_i introduced in Chapter 11 in the form

$$E_{an}(\theta, \varphi) = K_1 \sin^2 \theta + K_2 \sin^4 \theta. \qquad (12.5.1)$$

In lowest order approximation, the anisotropy constants K_1 and K_2 are related to these crystal-field parameters A_n^m via the relations (Lindgard and Danielsen, 1975; Rudowicz, 1985):

$$K_1 = -\tfrac{3}{2} N_R B_2^0 \langle O_2^0 \rangle - 5 N_R B_4^0 \langle O_4^0 \rangle = -\tfrac{3}{2} \alpha_J \langle r^2 \rangle N_R A_2^0 \langle O_2^0 \rangle - 5\beta_J \langle r^4 \rangle N_R A_4^0 \langle O_4^0 \rangle, \tag{12.5.2}$$

$$K_2 = -\tfrac{35}{8} N_R B_4^0 \langle O_4^0 \rangle = -\tfrac{35}{8} \beta_J \langle r^4 \rangle N_R A_4^0 \langle O_4^0 \rangle. \tag{12.5.3}$$

Similar expressions can be derived for the higher order constants K_i. The quantities $\langle O_n^m \rangle$ are thermal averages of the Stevens operators O_n^m. For instance,

$$\langle O_2^0 \rangle = \langle 3 J_z^2 - J(J+1) \rangle. \tag{12.5.4}$$

These thermal averages or statistical averages can be obtained by calculating O_n^m for each of the $2J+1$ crystal-field-split states, multiplying with the probability that a given state is occupied at a given temperature and then summing over all $2J+1$ states. The procedure is similar to that used in Chapter 3 for calculating the thermal average of the magnetic moment by means of Eq. (3.1.4). In general, this requires considerable computational effort. For practical purposes, it is sometimes useful to bear in mind that the thermal averages $\langle O_n^m \rangle$ can be shown to vary with a high power of the reduced rare-earth-sublattice magnetization $m_R = M_R(T)/M_R(0)$:

$$\langle O_n^m \rangle_T = \langle O_n^m \rangle_0 [m_R(T)]^p \tag{12.5.5}$$

with $p = n(n+1)/2$. The second- and fourth-order terms therefore vary with temperature as m_R^3 and m_R^{10}, respectively. This means that at room temperature it is generally sufficient to consider only the second-order terms because the strong temperature dependence has made the fourth-order terms negligibly small. In this approximation, one has $K_2 \approx 0$ (Eq. 12.5.3) and in the expression for K_1 (Eq. 12.5.2) only the term with $\langle O_2^0 \rangle$ is retained. This means that if we would know the value of A_2^0 for a given compound, we would be able to obtain the sign and the approximate value of K_1 from Eq. (12.5.2) by using the data listed in Table 5.2.1.

Although it has not explicitly been mentioned in the discussion given in the preceding sections, it will be clear that one of the requirements for permanent-magnet materials is that the magnetization adopts a unique direction as can be realized in compounds having crystal symmetries lower than cubic. For the case mentioned above that only the lowest order term K_1 contributes to the anisotropy, one finds for the magnetization M in hexagonal or tetragonal compounds (see Chapter 11) that

$$M \parallel c \quad \text{for } K_1 > 0,$$
$$M \perp c \quad \text{for } K_1 < 0.$$

In the latter case, the magnetization vector may have any direction in a plane perpendicular to the c direction. In this plane, there is no anisotropy-energy barrier that prevents the magnetization after alignment by an external field from rotating into the opposite direction. The conclusion therefore is that compounds with $K_1 < 0$ are not suitable for application as permanent magnets. For rare-earth compounds of the type $R_2Fe_{14}B$, the second-order

SECTION 12.5. MAGNETIC ANISOTROPY AND EXCHANGE COUPLING

crystal-field parameter A_2^0 is positive. Using Eq. (12.5.2) and the data listed in Table 5.2.1, one finds

$$K_1 > 0 \quad \text{for } R = \text{Ce, Pr, Nd, Tb, Dy, and Ho;}$$
$$K_1 < 0 \quad \text{for } R = \text{Sm, Er, Tm, and Yb.}$$

This shows that only $R_2\text{Fe}_{14}\text{B}$ compounds with the former six elements can be regarded as suitable for permanent magnets. However, for strong magnets one needs a high magnetization. This is realized only if the moments of the R atoms are parallel to the moments of the Fe atoms, meaning that the R-sublattice magnetization must be coupled parallel to the Fe-sublattice magnetization. Numerous experimental investigations and band-structure calculations have shown that the magnetic-coupling constant describing the magnetic coupling between these two sublattices equals about $J_{R\text{Fe}} = -3 \times 10^{-22} J$. Substituting this value for J_{AB} in Eq. (4.4.9), together with $g_A = g_{\text{Fe}} = 2$ and the values listed for $g_B = g_J$ in Table 2.2.1, one finds that $N_{AB} = N_{R\text{Fe}} > 0$ for $R = $ Ce, Pr, and Nd, but that $N_{AB} = N_{R\text{Fe}} < 0$ for $R = $ Tb, Dy, and Ho. For $R_2\text{Fe}_{14}\text{B}$ compounds with the latter three elements, the moments of the two magnetic sublattices point in opposite directions and consequently lead to a total magnetization value which is too low for permanent magnets.

Let us now turn to the three remaining compounds $\text{Ce}_2\text{Fe}_{14}\text{B}$, $\text{Pr}_2\text{Fe}_{14}\text{B}$, and $\text{Nd}_2\text{Fe}_{14}\text{B}$. Cerium is known to have an unstable valence. The reason for this is that it has only one 4f electron in the trivalent state (see Table 2.1.1). In metallic systems, an electronic configuration of lower energy can often be reached when this electron is promoted to the conduction band, whereby the Ce ion adopts the tetravalent state. This usually happens when Ce is combined with 3d transition metals. For the Ce ion, the loss of its 4f electron implies the loss of its localized 4f moment and the corresponding rare-earth-ion anisotropy. The magnetic anisotropy in $\text{Ce}_2\text{Fe}_{14}\text{B}$ is therefore only due to the Fe sublattice, which is too small for permanent-magnet applications.

Most of the powerful modern permanent-magnet materials are based on $\text{Nd}_2\text{Fe}_{14}\text{B}$. The reason why $\text{Pr}_2\text{Fe}_{14}\text{B}$ has not qualified is not a physical one. The natural abundance of Pr is much lower than that of Nd which implies that the price of the former is higher than that of the latter and consequently hampers large-scale industrial applications of $\text{Pr}_2\text{Fe}_{14}\text{B}$.

The above discussion may have shown which arguments are behind the remarkable fact that for a given technological application only one out of the 15 available rare-earth elements qualifies. It is illustrative to compare this with the two other rare-earth permanent-magnet materials listed in Table 12.5.1.

Table 12.5.1. Comparison of the magnetic properties of several commercially available permanent-magnet materials

Material	T_C (°C)	$(BH)_{\max}$ (kJ m^{-3})	B_r (T)	dB_r/dT (%/degr.)	$_JH_c$ (kA m^{-1})	$_BH_c$ (kA m^{-1})
Ferroxdure (SrFe$_{12}$O$_{19}$)	450	28	0.39	−0.2	275	265
Alnico 4	850	72	1.04	−0.015	—	124
SmCo$_5$	720	130–180	0.8–0.9	−0.01	1100–1500	600–670
Sm(Co, Fe, Cu, Zr)$_7$	800	200–240	0.95–1.15	−0.03	600–1300	600–900
Nd–Fe–B (sintered magnet)	310	200–280	1.0–1.2	0.13	750–1500	600–850

It can be seen in Table 12.5.1 that the advantage of SmCo$_5$ and SmCo$_7$ over Nd$_2$Fe$_{14}$B is their much higher Curie temperature. These materials are preferred as permanent magnets in electrical machines having a high use temperature, as in several automotive and aircraft applications with use temperatures in the range 200–300°C. The reason why, of the RCo$_5$ and RCo$_7$ series of compounds, only the compounds with $R = $ Sm qualify for permanent-magnet materials can be sketched along the same lines as given above for R_2Fe$_{14}$B. Because J_{RCo} is also negative, high total magnetization values are only obtained when R belongs to the light rare-earth elements. In contradistinction to the crystal structure of the R_2Fe$_{14}$B compounds, one has for the RCo$_5$ and RCo$_7$ series that $A_2{}^0 < 0$. Using again Eq. (12.5.2) and the data listed in Table 5.2.1, one finds

$$K_1 < 0 \quad \text{for } R = \text{Ce, Pr, Nd, Tb, Dy, and Ho};$$
$$K_1 > 0 \quad \text{for } R = \text{Sm, Er, Tm, and Yb}.$$

This leaves $R = $ Sm as the only possible rare-earth element that can be used in RCo$_5$- and RCo$_7$-based permanent magnets.

At temperatures below room temperature it is no longer legitimate to ignore the fourth-order term in Eqs. (12.5.2) and (12.5.3). Although, in Nd$_2$Fe$_{14}$B, at room temperature the value of B_4^0 is only approximately 1% of B_2^0, the fourth-order term will dominate at low temperatures. This leads to a temperature dependence of the anisotropy constants K_1 and K_2 as shown in Fig. 12.5.1. It can be seen in this figure that K_1 changes sign at the spin-reorientation temperature $T_S = 130$ K. Below this temperature, the preferred magnetization direction starts to deviate from the c direction and for each temperature has a direction given by Eq. (11.4) introduced in Chapter 11:

$$\sin^2\theta(T) = -\frac{K_1(T)}{2K_2(T)}. \tag{12.5.6}$$

It can be seen in Fig. 12.5.2 that the tilt angle θ reaches about 30° at 4.2 K. The results shown in Fig. 12.5.2 make it also clear that permanent magnets based on Nd$_2$Fe$_{14}$B lose their usefulness at cryogenic temperatures.

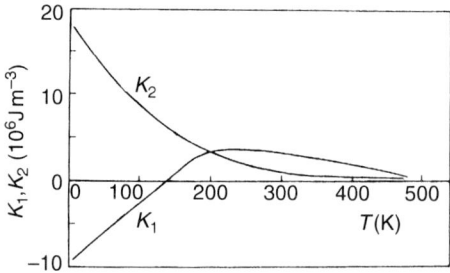

Figure 12.5.1. Temperature dependence of the anisotropy constants K_1 and K_2 in Nd$_2$Fe$_{14}$B.

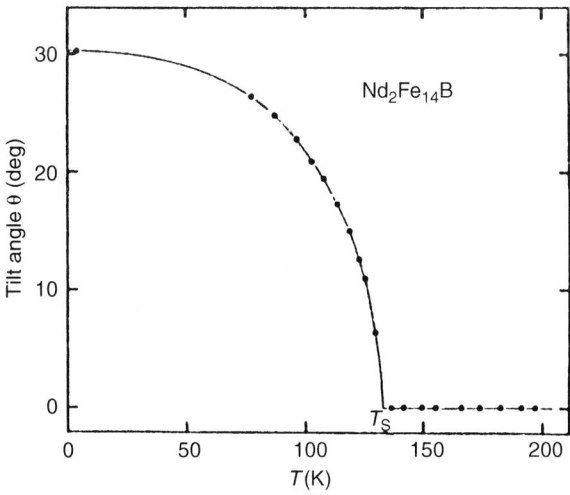

Figure 12.5.2. Temperature dependence of the tilt angle θ in $Nd_2Fe_{14}B$. After Tokuhara et al. (1995).

12.6. MANUFACTURING TECHNOLOGIES OF RARE-EARTH-BASED MAGNETS

The simplest production route for $Nd_2Fe_{14}B$ permanent magnets is schematically represented in Fig. 12.6.1. This is a well-established powder-metallurgical treatment that leads to high-performance magnet bodies. The manufacturing process to prepare $SmCo_5$-type permanent magnets is basically the same.

The main steps consist of alloy preparation, pre-milling, milling, control, and adjustment of the overall composition, particle alignment and pressing, sintering and heat treatment. After this treatment, the sintered magnet bodies can be machined into the shape desired, and are then magnetized. The consecutive steps shown in Fig 12.6.1 will be discussed in more detail below.

The most common way of alloy preparation is vacuum melting of the components in an induction furnace. First Fe and B are melted together in an alumina (Al_2O_3) crucible under purified argon gas. Subsequently, the reaction vessel is degassed under vacuum and Nd metal is added to the melt after the latter has reached a temperature only slightly above the Fe–B liquidus temperature. The casting is done in such a way so as to allow rapid cooling of the melt in order to prevent oxidation as far as possible. The composition of the alloy is generally chosen somewhat more Nd-rich than would correspond to the formula composition $Nd_2Fe_{14}B$. In that case, the $Nd_2Fe_{14}B$ grains are surrounded by small amounts of the Nd-rich eutectic present in the Nd–Fe–B phase diagram. The presence of intergranular material of the eutectic composition is important in the liquid-phase sintering process.

The chill-cast alloys are generally obtained in the form of large ingot lumps, too large for direct milling. These lumps are therefore first crushed by means of hammer mills. After a sufficiently small particle size has been reached, further size reduction is achieved

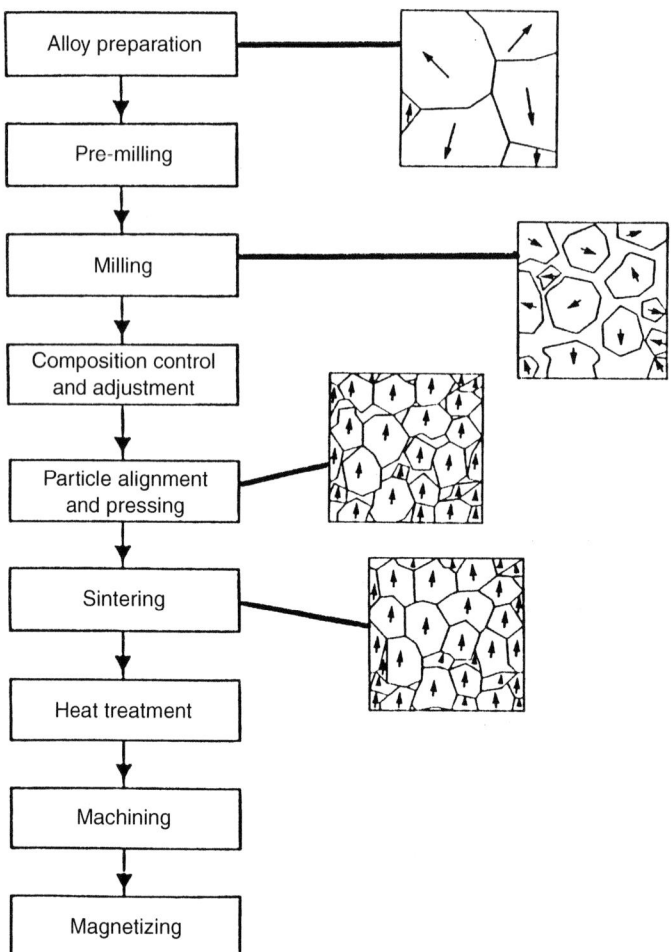

Figure 12.6.1. Consecutive steps in the manufacturing of permanent magnets based on $Nd_2Fe_{14}B$.

by means of ball milling, mortar grinding, or jet milling. In order to avoid oxidation, all milling operations have to be performed in an inert-gas atmosphere. Organic liquids frequently used for ball milling and attrition milling are freon, cyclohexane, or toluene. In all these cases, one has to be aware of a considerable explosion hazard because of the presence of the Nd-rich eutectic in the cast alloy (i.e., the presence of fine Nd particles in the coarse powder). After the milling treatment, the powder is dried under vacuum or by gentle heating in a flow of purified-argon gas.

The purpose of the milling process is to obtain fine particles that, in the most favorable cases, can be regarded as small single crystals. Because grain boundaries are absent, one may expect that the particles will have only one single axis of preferred magnetization. This

SECTION 12.6. MANUFACTURING TECHNOLOGIES OF RARE-EARTH-BASED MAGNETS

offers the possibility to reach almost perfect particle alignment in the alignment step (see also Fig. 12.6.1). Numerous investigations have shown that the ultimate sintered magnets have a sufficiently high coercivity only if the $Nd_2Fe_{14}B$ particles are present in sufficiently small grains. The coercivity in $Nd_2Fe_{14}B$-type permanent magnets is nucleation controlled (see Section 12.4). One of the reasons why small particle sizes favor high coercivities is that the smaller the particle the lower the probability that it will contain an imperfection acting as nucleation center. It can be never completely avoided that some of the fine particles present contain nucleation centers and hence a fraction of the particles will be prone to magnetization reversal in a demagnetizing field. The magnetization reversal will affect, however, only the latter particles and not spread into the whole magnet body. The overall magnetization reversal will therefore be very modest for small particle sizes and may even remain unnoticed.

In order to obtain an anisotropic magnet with the highest possible magnetization in a given direction, the powder particles have to be aligned after milling by means of an external magnetic field. After the magnetic alignment, the powder is pressed isostatically to yield a compact powder that, after sintering, has a sufficiently high density. It is commonly assumed that the degree of particle alignment does not change during isostatic pressing. Generally speaking, it is desirable to apply a high compacting pressure, but this pressure should not be chosen too high because it may then cause severe particle misorientation. Particle alignment and pressing can also be performed simultaneously. A non-magnetic die is used in this case, the desired magnetization direction being determined by the direction of the magnetic field set up in the cavity of the die.

The sintering step is essential for attaining high values of the ultimate magnetization and coercivity. Isostatic pressing or die-pressing alone is known to lead to densities of only 80% of the theoretical density. Liquid-phase sintering leads to much higher densities, up to 99% of the theoretical density. In that case, the overall composition of the alloy is chosen in such a way that after casting, small amounts of a low-melting alloy component are present. Sintering is then performed at a temperature low enough for the main phase ($Nd_2Fe_{14}B$) to remain solid. Only the second phase melts and makes mass transport possible during sintering with the ultimate result that all voids disappear and all $Nd_2Fe_{14}B$ grains are surrounded by a thin layer of the low-melting intergranular material. At room temperature and above, the intergranular material is non-magnetic. It magnetically isolates the $Nd_2Fe_{14}B$ grains and prevents magnetization reversal to spread into the whole magnet body if in one (or more) of the grains a domain wall is nucleated in a demagnetizing field. We mentioned already that the presence of very small, magnetically well-isolated, particles is important for achieving high coercivity. The liquid-phase sintering has a second equally important advantage. The disappearance of voids and the concomitant high density implies a high magnetization per unit volume or per unit mass. This is of prime importance for the manufacture of magnets with high energy products because, as we showed already in Section 12.3, the energy product of (ideal) permanent magnets is proportional to the magnetization squared. A third advantage of the liquid-phase sintering is the absence of porosity in the ultimate magnet body, making it more resistant to corrosion and giving it a substantially higher mechanical strength than would have been obtained by pressing alone. More sophisticated manufacturing routes have been reviewed by Buschow (1998).

12.7. HARD FERRITES

The ferrites used for permanent-magnet purposes are the hexaferrites, also called hard ferrites or M-type ferrites. These are hexagonal compounds of the general formula $MeFe_{12}O_{19}$ with $Me = Ba$, Sr, or Pb which owe their hard-magnetic properties to their comparatively large magnetocrystalline anisotropy. The hard ferrites play a dominant role on the permanent-magnet market, which is mainly due to the low price per unit of available energy, the wide availability of the raw materials, and the high chemical stability.

The M ferrites crystallize in the magnetoplumbite structure, characterized by a close packing of oxygen and Me ions with Fe atoms at the interstitial positions. There are five of such interstitial positions, meaning that the magnetically ordered structure is composed of five different magnetic sublattices.

Each of the Fe^{3+} ions in $BaFe_{12}O_{19}$ carries a magnetic moment of 5 μ_B. The moment of the Fe ions residing on the same crystallographic position are ferromagnetically aligned but the coupling between Fe moments at the different crystallographic positions may be ferromagnetic as well as antiferromagnetic. All these couplings are determined by the so-called superexchange interaction, mediated by the O atoms. There is a strong preference for ferromagnetic coupling when the angle Fe–O–Fe approaches 180° and the distance Fe–O–Fe becomes smaller. This is the reason why the magnetic moments of the five Fe sublattices are not mutually parallel. Two of the Fe sublattices have their moments oriented antiparallel to those of the other three. This ferrimagnetic arrangement of the resultant spin structure leads to a net moment per unit cell of only 40 μ_B (at 4.2 K). More details regarding the superexchange interaction can be found in the review of Guillot (1994).

The Curie temperatures of the $MeFe_{12}O_{19}$ compounds are fairly high and equal to 740, 750, and 725 K for $Me = Ba$, Sr, and Pb, respectively. Of particular interest in the M ferrites is the temperature dependence of the saturation polarization J_s. Results for the Ba compound are shown in Fig. 12.7.1. It may be inferred from this figure that between T_C and room temperature, the J_s values increase with decreasing temperature much more slowly than would be expected on the basis of the Brillouin function. As a consequence, these materials have a relatively low value of J_s at room temperature (much lower than the value corresponding to the saturation moment of 40 μ_B per formula unit mentioned above) while the temperature coefficient of J_s is fairly high ($-0.2\%\,K^{-1}$).

The magnetocrystalline anisotropy in the M ferrites is generally considered as arising from spin–orbit coupling. It is characterized by a comparatively high positive value of the anisotropy constant K_1 while higher order constants (K_2, K_3, ...) are negligibly small. This situation corresponds to an easy magnetization direction along the c-axis.

The temperature dependence of K_1 for $BaFe_{12}O_{19}$ is shown in Fig. 12.7.1, together with the temperature dependence of the anisotropy field $H_{an} = 2K_1/J_s$. Because J_s decreases more strongly with temperature than K_1 in the lower temperature range, one finds that H_{an} first slightly increases with temperature before it eventually decreases. This is a rather unusual behavior of the anisotropy field. It leads to an unusual behavior also for the coercivity. In magnets made of hard ferrites, the coercivity increases when the temperature is raised above room temperature whereas it decreases in all other permanent-magnet materials known. This is an advantageous property for high-temperature applications of such magnets.

SECTION 12.7. HARD FERRITES

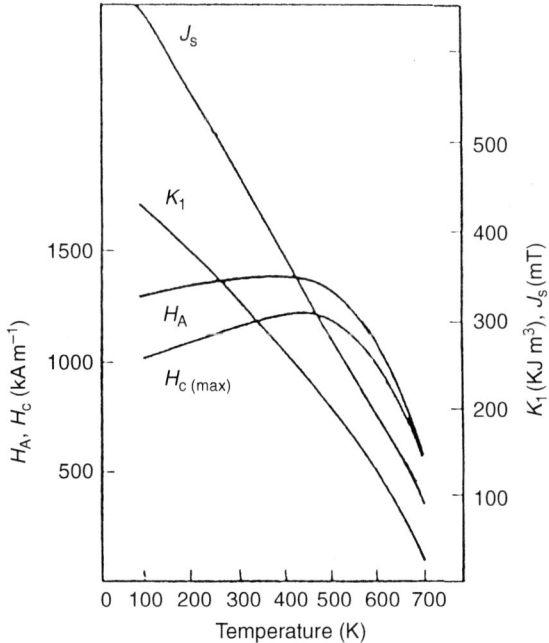

Figure 12.7.1. Temperature dependence of J_s, K_1, and H_{an} in $BaFe_{12}O_{19}$. After Kools (1986).

Solid bodies of permanent magnets based on M ferrites can be made either by sintering or by plastic bonding. Of both types of magnets, anisotropic as well as isotropic forms are applied. The former types are characterized by higher remanences owing to the magnetically induced alignment of the powder particles during processing. In the latter magnets, the orientation of the powder particles has a random distribution. The anisotropic sintered form is the most important.

The starting materials for the preparation of $MeFe_{12}O_{19}$ are $MeCO_3$, Fe_2O_3, and possible additives (SiO_2 and/or B_2O_3). Appropriate amounts of powders of these materials are dry-mixed and the resulting mixture is subsequently granulated in a disc pelletizer to granules of roughly 5 mm diameter. During the pre-firing process, performed at about 1250°C in air, the raw materials react to form the compound $MeFe_{12}O_{19}$. Comminution of the hard pre-fired granules is achieved by wet milling with steel balls. This latter process leads to a thick suspension (slurry) in which the fine powder particles (preferably single crystals) have sufficient mobility to align themselves along the preferred magnetization direction when an external field is applied during wet pressing. The resultant compacts are first dried and then sintered in air at about 1250°C. Anisotropic shrinkage occurs during the sintering process. For this reason, the pole faces of the sintered bodies have to be ground afterwards when accurate dimensional control of the magnets is required. For more details regarding processing, the reader is referred to the reviews by Kools (1986) and McCaig and Clegg (1987).

Permanent magnets made of hard ferrites can be characterized as low-cost low-performance magnets. Their application is widespread, main applications being anisotropic segments for electric motors, anisotropic rings for loudspeakers, and large anisotropic blocks for ore separators. Applications in which the use temperature can become substantially higher than room temperature may profit from the fact that the ferrite magnets have a high chemical stability and that the coercivity increases rather than decreases with temperature. A distinct disadvantage of magnets made of hard ferrites is their low $(BH)_{\max}$ value (see Table 12.5.1). It requires generally magnets of comparatively large size and this restricts their application to magnetic devices in which weight and space are not a concern.

12.8. ALNICO MAGNETS

Nowadays, the Alnico alloys have become a less important group of permanent-magnet materials. They contain Fe, Co, Ni, and Al with small amounts of Cu and Ti as additives. The Alnicos, like the sintered $R_2Fe_{14}B$- and $SmCo_5$-type magnets discussed in Section 12.5, are fine-particle magnets, consisting of ferromagnetic particles in a non-magnetic matrix. However, there is an important difference where the rare-earth-based magnets are concerned. In the Alnico alloys, the fine-particle structure is not the result of powder metallurgy but the result of a metallurgical precipitation reaction that takes place in the solidified ingots of the alloy.

The important role played by the microstructure of Alnico alloys is most conveniently discussed by means of alloys of the composition Fe_2NiAl, although Alnico alloys have in general a much more complicated composition, including Co. The pseudobinary section FeNiAl in the phase diagram is shown in Fig. 12.8.1. Permanent-magnet alloys close in composition to Fe_2NiAl are commonly prepared by a homogenization treatment at 1250°C. It can be seen in Fig. 12.8.1 that the alloy consists of one single phase (α) at this temperature. However, at lower temperatures there is a miscibility gap. The presence of this miscibility gap causes the α phase to decompose into two different phases α_1 and α_2 when an alloy of a composition falling into the gap region is kept at temperatures confined within the gap for some time. This second heat treatment is of prime importance for the formation of a microstructure in the alloy ingot that gives it the desired hard-magnetic properties.

Both the Fe-rich particles (α_1 phase) and the non-ferromagnetic or weakly ferromagnetic NiAl-rich matrix (α_2 phase) have the bcc structure. This circumstance is one of the reasons that the phase separation of α into α_1 and α_2 upon annealing at a temperature within the gap proceeds by so-called spinodal decomposition rather than by the normal nucleation and growth process. This has important consequences for the microstructure and the magnetic properties of the alloys, as will be discussed below.

Although the decomposition proceeds spontaneously, the rate of the spinodal decomposition of the α phase into α_1 and α_2 is diffusion limited. This means that the decomposition process will reach completion within a reasonable time only if the atoms are able to diffuse to a sufficient extent. Atomic motion during diffusion requires an activation energy that can be supplied only if the temperature is sufficiently high. Therefore, the decomposition rate is sufficiently high only at relatively high temperatures (850°C). The nature of the spinodal-decomposition process is such that the concentrations of the Fe atoms in the two phases show a periodic variation (sinusoidal) and the amplitude of the composition fluctuations increases with time until the phase separation into α_1 and α_2 is complete. The whole

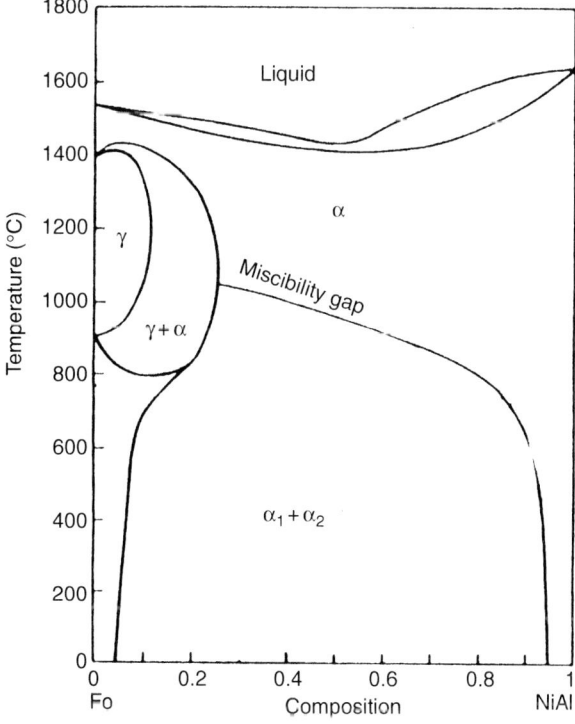

Figure 12.8.1. Pseudobinary section of the Fe–Ni–Al phase diagram. After Marcon et al. (1978).

process takes only a very short time when performed at 850°C, being completed in a matter of seconds or minutes. At lower temperatures, longer times are required.

The formation and growth of the particles to their final shape and size occurs almost entirely during the spinodal-decomposition reaction at 850°C. The driving force for this reaction is the reduction of the interfacial energy between the particles (α_1) and the matrix (α_2). Although the interfacial energy is small (10^{-3}–10^{-1} J m^{-2}), it is sufficiently large to promote particle growth.

In general, a third heat treatment is required for good quality magnets. The main effect of this third heat treatment, performed at 600°C, is to increase the difference between the saturation polarization of the Fe-rich particles and the surrounding matrix (NiAl-rich). During annealing, a continuous change in the composition takes place due to the diffusion of Fe atoms to the ferromagnetic particles.

The magnetic anisotropy of Alnico magnets is due to the ferromagnetic particles. However, this is not the type of magnetocrystalline anisotropy as found, for instance, in rare-earth-based magnets (see Section 12.4). In Alnico alloys, the magnetic anisotropy is due to the rod-like shape of the ferromagnetic particles. It originates from the difference in demagnetizing factor (see Chapter 8) in the two extremal directions of the particles, which requires that the easy magnetization direction is along the long direction of the rod-shaped particles. The spinodal decomposition alone does not produce a sufficiently large

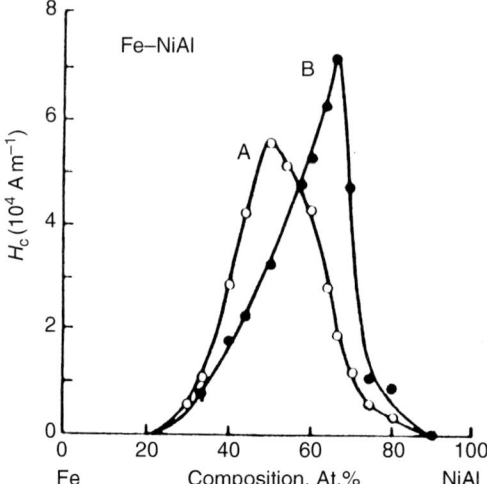

Figure 12.8.2. Dependence of the coercivity in Fe–NiAl alloys on composition and on heat treatment. The results shown by curve A were obtained after quenching of the ingot followed by an optimal heat treatment. The results of curve B refer to an alloy that was continuously cooled after casting. After de Vos (1969).

shape anisotropy in the primary ferromagnetic α_1-phase particles. Since the difference in saturation magnetization of the α_1 particles and the matrix is relatively small, the effective shape-anisotropy field of the particles is also small in spite of the elongation. Therefore, the heat treatment at 600°C is desirable in order to increase the difference in magnetization and the concomitant shape anisotropy. This, in turn, makes it possible to obtain the highest coercivities and the optimal permanent-magnet properties. The tempering treatment at about 600°C usually takes several hours.

In Fig. 12.8.2, a few examples are shown of how the intrinsic coercivity (H_c) for Fe–NiAl alloys can be varied with composition and heat treatment. In case A, the coercivities were determined after quenching and after tempering treatments to give the optimal coercivity. The results displayed by curve B were obtained by means of the more attractive manufacturing route of continuously controlled cooling of these materials. It should, however, be borne in mind that the overall magnetization of the magnets decreases with decreasing Fe content. Therefore, the compositions corresponding to the maximum coercivity need not necessarily correspond to the optimal composition of the ultimate magnet.

The interfacial energy responsible for the growth of the ferromagnetic particles depends on the crystallographic orientation of the boundary between the α_1 and the α_2 phase. Therefore, the particle growth is anisotropic, which results in an elongation parallel to the $\langle 100 \rangle$ directions of the cubic alloy. Significant improvements of the magnetic properties are therefore generally obtained by controlled cooling of the alloys from 1200°C to about 800°C in a saturating magnetic field. This thermomagnetic treatment leads to anisotropic magnets in which the easy magnetization direction of the grains formed during the spinodal decomposition is parallel to the direction of the magnetic field applied during cooling. The elongation

SECTION 12.8. ALNICO MAGNETS

of the particles in the field direction occurs because the magnetic free energy of the particles is lower when the axis with the lowest demagnetizing factor is in the direction of the applied field. In principle, one would obtain the best properties when the magnetic field is applied parallel to one of the three ⟨100⟩ directions (for instance [001]) of an oriented single crystal. In practice, instead of expensive single crystals, so-called columnar-crystallized Alnico alloys are applied. These alloys can be obtained by a grain-orienting process before the thermomagnetic and tempering treatment. It is achieved by casting the alloys in heated molds onto water-cooled steel or copper slabs. When the alloys solidify on the cold surface, the grains tend to grow with their long axis parallel to the ⟨100⟩ directions, perpendicular to the cold surface. The result is then a semicolumnar alloy in which the columnar axis is parallel to one of the ⟨100⟩ directions, for instance, parallel to [001]. These alloys are often referred to as Alnico directed grain (DG). The thermomagnetic treatment is subsequently applied with the magnetic field parallel to the [001] direction of the latter alloys.

As already mentioned above, the magnetic properties in the easy magnetization direction can be further improved by subsequent tempering for several hours at about 600°C. The purpose of the 600°C tempering treatment is to enhance the difference in magnetic polarization between the α_1 and α_2 phases by diffusion of magnetic atoms from α_2 to α_1 and of non-magnetic atoms from α_1 to α_2. An example of a microstructure, observed by electron microscopy, of a grain-oriented alloy after thermomagnetic and tempering treatment is shown in Fig. 12.8.3. The direction of the magnetic field applied during the thermomagnetic treatment corresponds to the elongated direction of the columnar particles in Fig. 12.8.3a.

The relatively high coercivities and remanences in the Alnicos are principally due to shape anisotropy of elongated Fe rich particles in a non-ferromagnetic matrix. The Stoner–Wohlfarth theory, already mentioned in Section 12.4, predicts that the coercivity in these materials is proportional to the saturation polarization J_s of the Fe-rich particles and to a factor related to the difference in the effective demagnetization factors perpendicular ($N_\perp$) and parallel ($N_\parallel$) to the preferred direction of magnetization in the particles. Using Eq. (12.5.2) and bearing in mind that K_1 in these materials is negligibly small, one finds that

$$H_c = \frac{1}{\mu_0} f(Q)[N_\perp - N_\parallel] J_s. \tag{12.8.1}$$

Here, $f(Q)$ is an averaging factor that takes account of the various orientations of the preferred axes of the particles with respect to the direction in which H_c is measured. If one assumes that the particles are magnetically non-interacting uniaxial single-domain particles arranged at random, the factor $f(Q)$ equals about 0.5. But $f(Q)$ may approach the value one in highly elongated particles. In the case of spheroid particles, there is a considerable difference in the demagnetizing factor for particles magnetized perpendicular and parallel to the flat surface of the spheroid. In the limit of an extremely flat and elongated spheroid, one has $N_\perp - N_\parallel = 1 - 0 = 1$. Hence, the coercivity in such materials may reach an upper limit, according to Eq. (12.8.1), equal to $H_c = J_s/\mu_0 = M_s$. For $M_s = 1.7$ MA m^{-1}, this upper limit becomes $H_c = 1.7$ MA m^{-1} and, for $N_\perp - N_\parallel = 0.5$, the coercivity becomes 850 kA m^{-1}. Actual values found in Alnico materials are much lower, as may be seen from the data shown in Fig. 12.8.2. This has primarily been attributed to the less perfect shape of the thin ferromagnetic particles and also to the fact that the α_2 phase is magnetic to some extent.

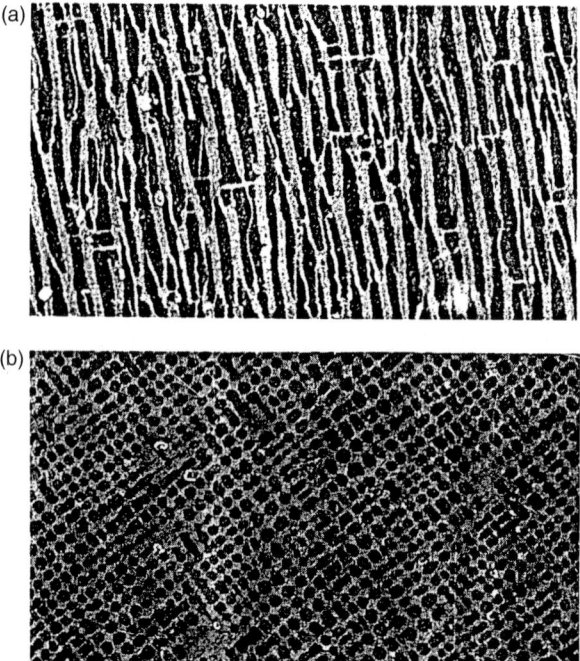

Figure 12.8.3. Microstructure of a "Ticonal XX" or grain-oriented magnet alloy as observed in an electron microscope. The microstructure consists of a distribution of fine magnetic needles (mainly FeCo) in a matrix (NiAl) that has a much smaller magnetic moment. (a): Plane of observation parallel to the preferred direction of magnetization. (b): The same, but perpendicular to the preferred direction. After de Vos (1969). Magnification 50,000.

A most favorable magnetic property of the Alnico alloys is that these materials have very high Curie temperatures (700–850°C) which leads to excellent flux stability at elevated temperatures. The Alnico alloys are chemically and metallurgically very stable. In fact, Alnico is the only magnet material that has some long-term utility at temperatures up to 500°C. A drawback of the Alnico alloys is that their coercivity is low in comparison to the rare-earth-based magnets described in Section 12.5. The non-linear behavior of the $B(H)$ curve in the second quadrant (Fig. 12.2) is a serious disadvantage in device design and dynamic operation, and this limits the attainable energy product in spite of the high remanence.

References

Buschow, K. H. J. (1988) in W. P. Wohlfarth and K. H. J. Buschow (Eds) *Ferromagnetic materials*, Vol. 4, Amsterdam: North Holland Publ. Co. p. 1.

Buschow, K. H. J. (1998) in K. H. J. Buschow (Ed.) *Handbook of magnetic materials*, Amsterdam: North Holland Publ. Co., Vol. 10, p. 436.

Givord, D., Lu, O., Rossignol, M. F., Tenaud, P., and Viadieu, T. (1990) *J. Magn. Magn. Mater., 83*, 183.

Guillot, M. (1994) in R.W. Cahn et al. (Eds) *Materials science and technology*, Weinheim: VCH, Vol. 3B, p. 1.

SECTION 12.8. ALNICO MAGNETS

Kools, F. (1986) in M. B. Bever (Ed.) *Encyclopedia of materials science and engineering*, Oxford: Pergamon Press, *Vol. 4*, p. 2082.

Kronmüller, H., Durst, K. D., and Sagawa, M. (1988) *J. Magn. Magn. Mater.*, *74*, 291.

Kronmüller, K. (1991) in G. J. Long and F. Grandjean (Eds) *Supermagnets, hard magnetic materials*, Dordrecht: Kluwer Academic Publ., NATO ASI Series E, *Vol. 331*, p. 461.

Lindgard, P. A. and Danielsen, O. (1975) *Phys. Rev.*, *B11*, 351.

Marcon, G., Peffen, R., and Lemaire, H. (1978) *IEEE Trans. Magn.*, *14*, 685.

McCaig, M. and Clegg, A. G. (1987) *Permanent magnets in theory and practice*, London: Pentech Press.

Rudowicz, C. (1985) *J. Phys.*, *C18*, 1415.

de Vos, K. J. (1969) in A. E. Berkowitz and E. Kneller (Eds) *Magnetism and metallurgy*, New York: Academic Press *Vol. 1*, p. 473.

Tokuhara, K., Ohtsu, Y., Ono, F., Yamada, O., Sagawa, M., and Matsuura, Y. (1985) *Solid State Comm.*, *56*, 333.

Zijlstra, H. (1982) in E. P. Wohlfarth (Ed.) *Ferromagnetic materials*, Amsterdam: North Holland Publ. Co., *Vol. 3*, p. 37.

13

High-Density Recording Materials

13.1. INTRODUCTION

Several magnetic materials find an application as thin magnetic films in high-density recording devices. The most common methods for high-density recording are schematically represented in Fig. 13.1.1.

Case A represents conventional optical recording applied, for instance, in compact disks. A pattern of pits is burned by means of a pulsed laser beam into a non-magnetic film at the surface of the disk. This pattern is read out by means of a laser of much lower intensity, the reflected beam being out or in focus when it hits a pit or does not hit a pit, respectively.

Case B is a modification of case A. Local heating of a suitable surface layer by means of a pulsed laser beam leads to temporary local melting. After irradiation the melt cools at a sufficiently high rate to produce the amorphous state. The irradiated spot can be distinguished from the unmodified matrix by means of its lower reflectivity or larger transmittance. The advantage of case B compared to case A is the erasability of the written information. Instead of amorphization, one may also use local color changes produced by irradiation in special materials.

Case C illustrates the principles involved in magneto-optical recording. The thin surface layer of the disk consists, for instance, of an amorphous Gd–Fe or Gd–Co alloy. The requirements for alloys to be used as magneto-optical recording media include the following:

(i) An easy magnetization direction perpendicular to the film plane. The corresponding anisotropy of the material has to be sufficiently strong to overcompensate the tendency of the magnetization vector to lie in the film plane because of the lower demagnetizing factor in this direction.
(ii) A sufficiently high coercivity at room temperature but decreasing strongly with temperature.
(iii) A low thermal conductivity.

These properties offer the possibility of thermomagnetic writing of bits. This is accomplished by switching the magnetization direction of a tiny spot by local heating with a pulsed laser beam. The local heating brings the material in the spot area into a temperature range where the coercivity is low. The coercivity has to be low enough for the local demagnetizing

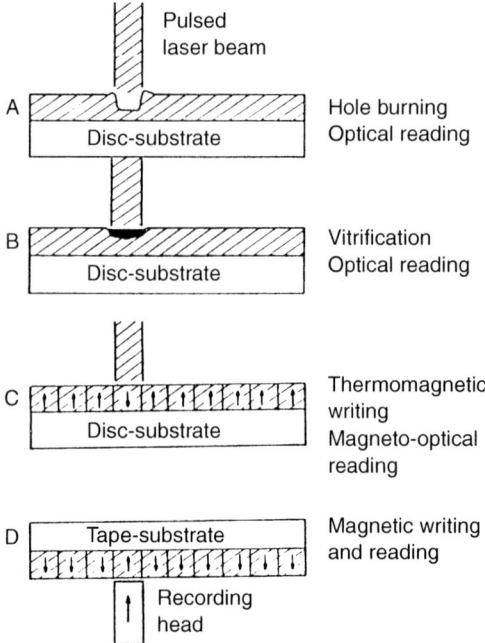

Figure 13.1.1. Schematic representation of various methods used for high-density recording.

field to reverse the magnetization of the irradiated material. After the laser beam has moved away, the temperature of the material returns to room temperature keeping its reversed magnetization direction. The written bits are stable at room temperature because the high room-temperature coercivity prevents further changes of the magnetization direction.

Properties (ii) and (iii) guarantee a comparatively small diameter of the reversed domain and therefore a large bit density. Reading of the written bits is performed by means of the Kerr effect, as will be discussed later. The advantage of magneto-optical recording compared to conventional optical recording (case A) is the erasability of the written information. By means of a special technique, that will not be discussed here, it is possible to overwrite old information by new one. A further advantage is the higher bit density.

Case D represents vertical recording by means of a recording head on a tape or a rigid disk covered with a thin magnetic film. In normal (longitudinal) magnetic recording, the easy magnetization direction is within the film plane. The difference between vertical recording and longitudinal recording and the requirement for the corresponding materials will be discussed together with the physical problems associated with their application.

It is good to bear in mind that magnetic and magneto-optical recording media fall in the class of hard-magnetic materials because of the requirement of sufficiently high coercivity that keeps the written artificial domain pattern from changing as a function of time. By contrast, inductive and magnetoresistive recording-head materials fall in the class of soft-magnetic materials. These materials and their application will be dealt with in Chapter 14.

13.2. MAGNETO-OPTICAL RECORDING MATERIALS

An important group of magneto-optical recording media is based on amorphous alloys of Gd-Fe or Gd-Co with some alloying additives to optimize the magnetic and magneto-optical properties.

The relative concentrations of rare-earth (R) and 3d elements are chosen in such a way that the R-sublattice magnetization exceeds the 3d-sublattice magnetization at low temperatures. The exchange-coupling constants J_{RFe} (J_{RCo}) responsible for the magnetic coupling between the R moments and the 3d moments are negative, as in the case of crystalline materials. The absolute values of these intersublattice-coupling constants are much smaller than the 3d-intrasublattice-coupling constant J_{FeFe} (J_{CoCo}). The R-intrasublattice-coupling constant J_{RR} is comparatively small and can be neglected in most cases. Using Eqs. (4.4.7) and (4.4.9) and the fact that J_{RFe} (J_{RCo}) < 0, one then finds that the 3d-sublattice moment is coupled antiparallel to the R-sublattice moment if the R component belongs to the heavy-R elements ($g_J > 1$; see Table 2.2.1). The temperature dependence of the magnetic polarization can then be calculated by means of Eqs. (4.4.15-4.4.19) and behaves as shown in Fig. 13.2.1.

It is essential for the application of amorphous R-3d alloys that their easy magnetization direction be perpendicular to the film plane, that is, $K_u > 0$. Here, we have used the symbol K_u instead of K_1 to indicate the difference from crystalline materials with uniaxial lattice symmetry.

Various models have been proposed in the literature that describe the origin of the positive anisotropy constant K_u found in some of the R-3d films. The model of

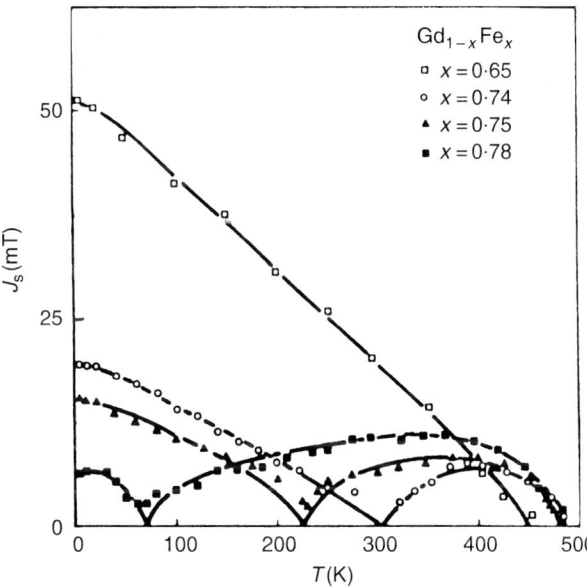

Figure 13.2.1. Temperature dependence of the magnetic polarization for several amorphous $Gd_{1-x}Fe_x$ alloys. After Hartmann (1982).

Gambino et al. (1973) is the most prominent one and will be discussed briefly. Gambino and co-workers conclude that short-range ordering of the atoms is the main source of anisotropy in sputtered Gd–Co films. These authors also provide a clue as to which type of short-range ordering causes the anisotropy. On the basis of studies on hcp-cobalt, they conclude that the easy magnetization direction most likely is due to the presence of Co–Co atom pairs having their pair axes perpendicular to this direction, the magnitude of the anisotropy energy being of the order of 10^{-22}–10^{-23} J per pair.

In order to understand the formation of such pair-atoms during vapor deposition, one has to consider the following. During the deposition process the ad-atom impinges on the film surface with considerable energy. After impingement, it rapidly loses this energy to the substrate and the main body of the film. If the substrate temperature is sufficiently high, the ad-atom will be able to move by means of surface diffusion to favorable sites of relatively low energy, so as to produce eventually a crystalline film. Low substrate temperatures and high evaporation rates do not favor such rearrangements of the ad-atoms and then may lead to amorphous films.

In the intermediate case, the ad-atom may still have the opportunity to jump to any of its nearest-neighbor surface sites, the jump probability being proportional to the corresponding activation energy. Differences in activation energy for jumps between the initial site and the nearest-neighbor surface site can have chemical, geometrical, and magnetic origins. This difference in activation energy for atomic jumps can be exploited for the generation of a higher concentration of Co pairs with their axes in the film plane than would correspond to a statistical distribution. Use is made of so-called bias sputtering, leading to conditions where an ad-atom bonded to a similar surface atom has a higher resputtering probability than an ad-atom bonded to a dissimilar atom. Consequently, there will be a greater statistical probability of Gd–Co pairs with their pair axes oriented perpendicular to the film plane than parallel to the film plane. The opposite holds for Co–Co pairs. This behavior of Gd–Co alloys is due to the fact that the bonding between a Co atom and a Gd atom is stronger than between two Co atoms or two Gd atoms. This is intimately related to the negative heats of solution of Gd in Co and of Co in Gd (see, for instance, de Boer et al., 1988). The corresponding heat-of-solution values are by far less negative in the case of Gd–Fe. The weaker bonding between Fe and Gd atoms is probably the reason why the perpendicular anisotropy is less easily attained by means of this method in the Gd–Fe alloys than in the Gd–Co alloys.

There are several observations that support the pair-ordering model of anisotropy. First, the anisotropy is relatively temperature independent near room temperature. The magnetic ordering of the Co sublattice is almost complete at room temperature, in contrast to the Gd sublattice that becomes magnetically ordered more gradually at lower temperatures. This indicates that the anisotropy is to be associated with the Co sublattice. Second, the growth-induced anisotropy increases with increasing resputtering but decreases at high deposition rates and low substrate temperatures.

Other models dealing with the occurrence of positive uniaxial anisotropy in amorphous R-3d alloys consider various types of shape anisotropy associated with structural inhomogeneities on a microstructural scale, including phase separation.

It was shown in Chapter 11 that in uniaxial materials, the following relation exists between the anisotropy field H_{an} and the anisotropy constant $K_1(K_u)$:

$$H_{an} = 2K_u/J_s. \tag{13.2.1}$$

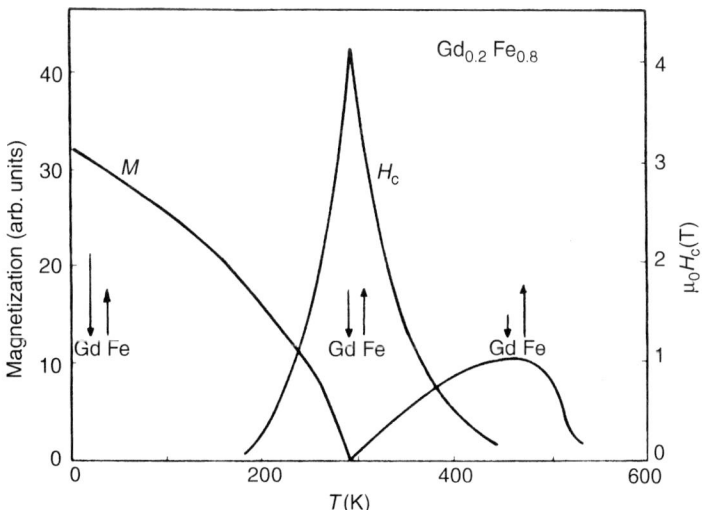

Figure 13.2.2. Schematic representation of the temperature dependence of the coercivity and the total magnetization of amorphous $Gd_{0.2}Fe_{0.8}$. The relative magnitude and orientation of the Gd- and Fe-sublattice magnetizations are indicated by arrows.

We mentioned already that the anisotropy constant K_u does not vary strongly at room temperature and below. However, J_s varies extremely strongly near the compensation temperature T_{comp}. In fact, since J_s becomes zero at T_{comp}, one expects that at the same temperature H_{an} will diverge. The coercivity is correlated with H_{an} so that it is plausible that the coercivity shows a very strong increase at T_{comp}. In practice, one observes a temperature dependence of the coercivity around the compensation temperature as shown in Fig. 13.2.2.

The strong temperature dependence of the coercivity is of prime importance for the writing of the domains with reversed magnetization direction. The local heating by means of a laser beam brings about a local reduction in coercivity so that the demagnetizing field can reverse the magnetization in the heated area. A strong decrease of the coercivity with respect to the room temperature value is most desirable because the temperature excursion ΔT needed to reverse the magnetization can be kept low and the same holds for the writing power of the laser beam. The temperature will again decrease quickly to room temperature after the laser beam has moved away. The original coercivity is restored and keeps the local magnetization in the opposite direction. Unlike an intermetallic R-3d compound of fixed composition, it is possible to vary the composition of an amorphous alloy continuously. This compositional freedom associated with the amorphous state makes it possible to choose the appropriate R/3d composition ratio in such a way that the maximum of the coercivity (occurring at T_{comp}) is located at a temperature close to room temperature.

Read-out of the written bits is done by means of a laser beam of lower intensity than the one used for writing the bits. It is essential for the read-out process that the laser beam be linearly polarized. In that case, the spots of reversed magnetization can be distinguished from regions of the original magnetization direction by means of the Kerr effect. In 1877, Kerr discovered that the plane of polarization of linearly polarized light is rotated over

Figure 13.2.3. Examples of magnetic domains written in a pre-grooved structure with a track spacing of 1.7 μm and made visible by means of a polarization microscope. The magnetic domains have lengths of about 2 μm. After Hartmann (1982).

a small angle φ_K when the light is reflected by a magnetic layer. This rotation of the polarization plane depends on the direction of the magnetization, that is, it is in opposite directions for regions having an opposite magnetization direction. The written bits can then be distinguished from the matrix region by means of Nichol prisms (or Mylar foils). An example of magnetic domains written and read-out using an amorphous Gd–Fe film is shown in Fig. 13.2.3.

If the substrate is translucent and the amorphous film is sufficiently thin, one may use transmitted, linearly polarized light to read out the written bits. Also, in this case there will be a rotation θ_F of the polarization plane (Faraday effect). The advantage of transmitted light is that the rotation angle θ_F increases with the thickness of the magnetic layer. This offers a better possibility of optimizing the contrast between written bits and the matrix, bearing in mind that the film is no longer translucent if it becomes too thick. A more detailed description of magneto-optical recording devices and materials can be found in the reviews of Buschow (1984), Reim and Schoenes (1990), and Hansen (1991).

It is interesting to discuss briefly the temperature dependence of φ_K or φ_F. Results obtained on several amorphous $Gd_{1-x}Fe_x$ films are shown in Fig. 13.2.4. These results have to be compared with the temperature dependence of the magnetization, shown for a number of such alloys in Fig. 13.2.1. It follows from the results of the latter figure that there is a compensation temperature in the temperature dependence of the magnetization of the amorphous $Gd_{1-x}Fe_x$ alloys when the Fe concentration falls into the range $0.66 \leq x \leq 0.78$. Inspection of the results shown in Fig. 13.2.4 makes it, however, clear that such features are absent in the temperature dependence of the Faraday rotation. This means that

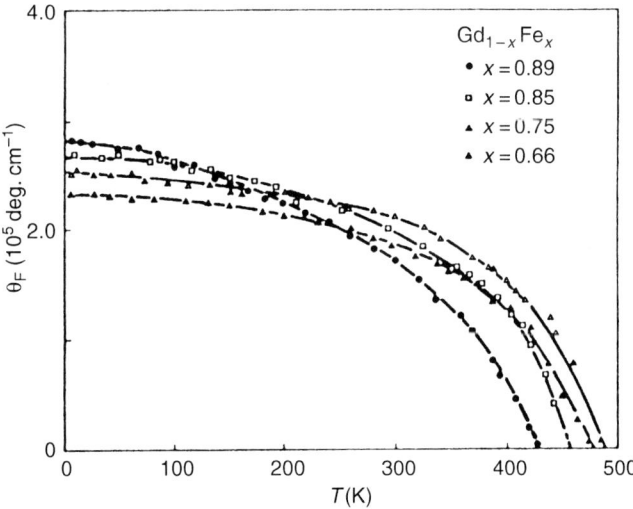

Figure 13.2.4. Temperature dependence of the specific Faraday rotation θ_F in several amorphous $Gd_{1-x}Fe_x$ alloys. After Hartmann (1982).

the magneto-optical rotation does not originate from the overall magnetization of the film but is due to one of the two sublattice magnetizations. This can be understood from the results shown in Fig. 4.5.1, illustrating that both sublattice magnetizations have a smooth temperature dependence even in a ferrimagnetic material with a compensation temperature.

In the upper part of Fig. 13.2.5, a schematic representation of the magnitude and direction of the two sublattice magnetizations around the compensation temperature is given. Here, we have assumed that the direction of the total magnetization $M = |M_{Fe} - M_{Gd}|$ follows the direction of the applied field, meaning that both the Fe-sublattice magnetization and the Gd-sublattice magnetization reverse their direction when passing from above T_{comp} to below T_{comp}. The Fe-sublattice magnetization is dominant in the high-temperature regime, whereas the Gd-sublattice magnetization dominates below the compensation temperature.

Hysteresis loops are shown for both temperature regions in the lower part of the figure. These results were obtained not by measuring the magnetization as a function of field strength but by measuring the rotation angle versus field strength. The fact that the hysteresis loop becomes reversed when passing the compensation temperature agrees with the notion that the optical rotation originates from only one of the two sublattice magnetizations and not from the total magnetization.

At this stage, it is difficult to decide which of the two sublattice magnetizations is responsible for the magneto-optical rotation, since both sublattice magnetizations change their direction when passing the compensation temperature. This dilemma has been solved by measuring the optical rotation at a fixed temperature on alloys of increasing Fe concentration. Results of magneto-optical measurements are shown in Fig. 13.2.6. It can be seen that the Kerr rotation φ_K (full curve) does not follow the total magnetization (broken curve), but increases with Fe concentration. This shows that the magneto-optical rotation

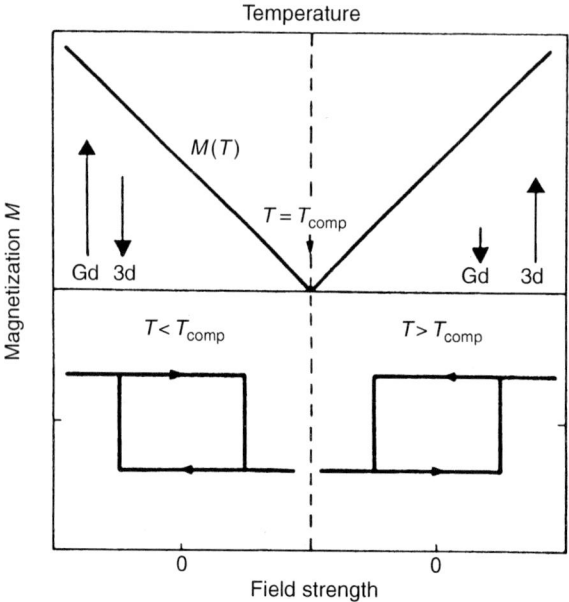

Figure 13.2.5. Top part: Temperature dependence of the total magnetization M around the compensation temperature T_{comp}. The relative magnitude and orientation of the Gd- and Fe-sublattice magnetizations are indicated by arrows. Bottom part: Schematic representation of magneto-optically measured hysteresis loops above and below T_{comp}.

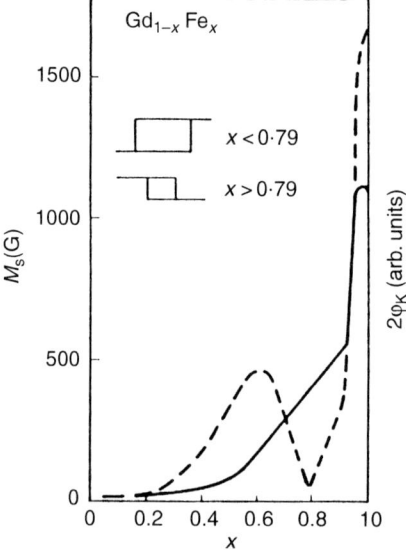

Figure 13.2.6. Concentration dependence of the saturation magnetization (broken curve) and Kerr rotation (full curve) in amorphous Gd_xFe_{1-x} films. Also shown are Kerr hysteresis loops obtained on both sides of the compensation composition ($x \approx 0.79$). After Imamura and Mimura (1976).

is due to the Fe sublattice. It can also be seen in the figure that there is a reversal of the hysteresis loops when going from the Gd-dominated range ($x < 0.79$) to the Fe-dominated range ($x > 0.79$).

13.3. MATERIALS FOR HIGH-DENSITY MAGNETIC RECORDING

Magnetic recording has been a subject of interest already for a long time. It has received additional impetus with the advent of computer systems and the associated demand for high-density recording devices. In most of such devices, digital magnetic recording is used in which a transducing head (write/read head) magnetizes small areas on a magnetic-recording medium so as to record digital data and scan the magnetized areas to read the data. The only commercially useful systems employed in the past were so-called longitudinal magnetic-recording materials having an easy axis of magnetization parallel to a major surface of the material.

For longitudinal magnetic recording, a head of the granular type is used. It comprises a core of a magnetically highly permeable material (see also Chapter 14), provided with a narrow air gap. The gap is placed transversely to the direction of movement of the magnetic-recording medium in such a way that flux coupling is possible. A current pulse applied to a coil wound around the core generates magnetic flux lines in the core which close along a path that comprises one edge of the gap, the part of the magnetic tape adjoining the gap, and the other edge of the gap. The flux passing through the magnetic layer in this manner causes data to be recorded. The data are read as the magnetized area on the medium moves past the gap, thereby closing the flux through the core. As a result, flux lines pass through the coil and induce an electric signal which is representative of the stored information.

The disadvantage of conventional longitudinal recording is that the system can handle only a rather restricted linear bit density. This restriction occurs because the magnetized areas in the magnetic layer are magnetically oriented in the longitudinal direction of the medium, that is, in the plane of the tape or the rigid disk. In conventional longitudinal recording methods, there is a certain maximum tolerable demagnetization field at the bit boundary, as a result of which the number of bits that can be stored per centimeter of the information track is limited.

A further problem arises when high recording currents are used. In that case, the magnetization pattern recorded will have a shape such that the magnetic-flux lines close inside the medium, which reduces the flux available for read out. Such a circular magnetization mode is schematically represented in Fig. 13.3.1. In order to obtain high densities, it is essential to avoid the nucleation of such magnetization modes. There are two methods to accomplish this. One is the use of longitudinal recording materials that have an enhanced longitudinal magnetization component. This can be achieved when the recording medium is made extremely thin so that the magnetization is forced to lie in the medium plane. The use of thin magnetic films is equivalent to media having a strong-shape anisotropy so that the magnetization is within the film plane. The thinner the film, the narrower the transition region will become. Such high density longitudinal recording media can be made from films consisting of chemically deposited Co–Ni–P or Co–P.

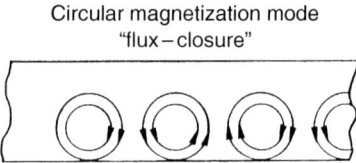

Figure 13.3.1. Circular magnetization mode existing in conventional longitudinal recording media.

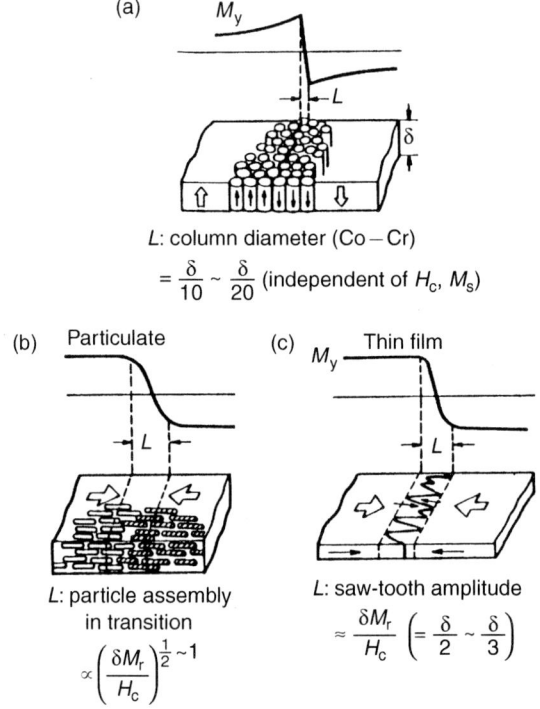

Figure 13.3.2. Model descriptions for transition regions in various types of magnetic recording media. (a) Perpendicular recording based on Co–Cr alloys, (b) longitudinal recording using particulate media, (c) longitudinal recording using thin films. After Suzuki (1984). ©1984 IEEE.

The second method is based on so-called perpendicular magnetic recording, in which materials are used that have an enhanced perpendicular magnetization component. These perpendicular recording materials have a high anisotropy. The preferred magnetization direction is perpendicular to the film plane, which inhibits the formation of the circular polarization mode. Thin films of Co–Cr alloys possess such favorable properties. They make it possible to obtain sharp transitions between domains of opposite magnetization, which is a prerequisite for high-density recording.

The two types of magnetic recording, longitudinal and perpendicular, are compared in Fig. 13.3.2. Step-like changes in the initial distribution of the magnetized areas in the medium would occur if the recording process were an ideal one. This is indicated in

SECTION 13.3. MATERIALS FOR HIGH-DENSITY MAGNETIC RECORDING

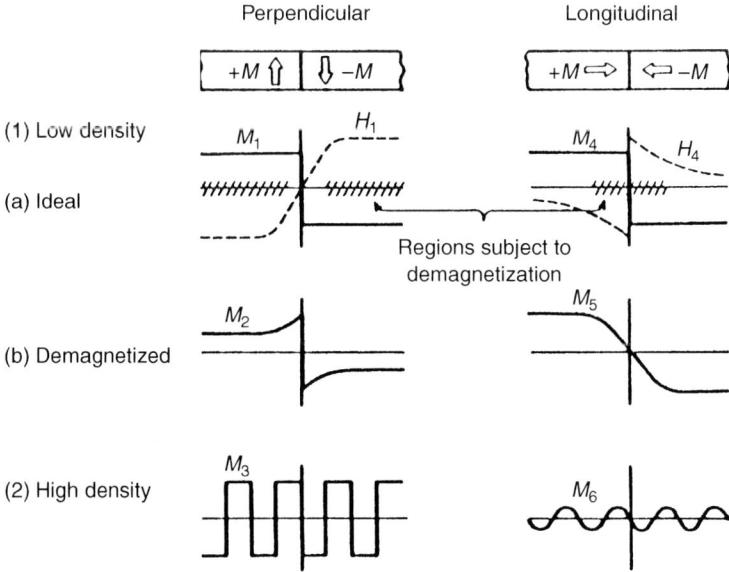

Figure 13.3.3. Simple model for magnetization transitions in perpendicular and longitudinal recording. In ideal recording processes, the initial magnetization transitions are given by step-wise changes. The top and middle part represent situations where the recording density is low and consecutive transitions are located far from each other. The lowest part of the figure shows the situation when there is a high density of domains and the corresponding transitions. After Suzuki (1984) ©1984 IEEE.

Fig. 13.3.3 by M_1 for perpendicular recording and by M_4 for longitudinal recording. However, the presence of demagnetizing fields (H_1 associated with M_1 and H_4 with M_4) makes the transition less sharp. In general, one may expect that demagnetization will occur in regions where the fields H_1 and H_4 are larger than the corresponding coercivities. It can be seen from the figure that there is hardly any demagnetization in the region around the transition center for perpendicular recording. Consequently, the transition (M_2) remains sharp. By contrast, the region around the transition for longitudinal recording is strongly demagnetized and leads to a broad transition (M_5).

It should be borne in mind that the explanations given above are based exclusively on the difference in magnetization direction in the two types of media. The sharp magnetization transition in perpendicular recording and the broad transition in longitudinal recording are therefore intimately connected with the intrinsic properties of the recording media, namely with their demagnetizing behavior. Models for the transition region and their sizes are shown for some typical recording media in Fig. 13.3.2.

In perpendicular recording, sputtered Co–Cr films are superior to many other perpendicular recording media, as regards perpendicular anisotropy, grain growth, and size. The films consist of tiny columns of hexagonal Co–Cr with their axes normal to the film plane. Each column is separated from the adjacent one by Cr-rich non-magnetic layers and therefore behaves as a magnetically isolated single-domain particle. It is mainly the shape anisotropy of each of the individual columns that gives rise to the perpendicular anisotropy.

The minimum magnetization transition length L for a Co–Cr film is assumed to be of the order of a column diameter, which is roughly one-tenth to one-twentieth of the film thickness, and is independent of the saturation magnetization M_s and coercivity H_c of the film. A possible magnetic-transition model for this film is shown in Fig. 13.3.2a.

In longitudinal recording, if conventional so-called particulate media are used, which consist of an assembly of coated magnetic particles (for instance, CrO_2 or γ-Fe_2O_3) dispersed in a binder, one expects a rather wide transition region as shown in Fig. 13.3.2b. The magnetization transition is composed of an assembly of particles in this case, and the transition width L is independent of the particle size. It can be shown that L is given by the expression (see, for instance, Mee and Daniel, 1987)

$$L = \frac{M_r \delta}{2\pi H_c}, \quad (13.3.1)$$

where M_r is the remanence, H_c the coercivity, and δ the film thickness. If one also takes into account the demagnetization in the write process, one finds a somewhat different value:

$$L = \sqrt{\frac{M_r \delta}{\pi H_c}}. \quad (13.3.2)$$

It follows from these expressions that these media must be made very thin if one wishes to obtain a high bit density. For particulate media, this requirement is difficult to achieve.

A better approach to high-density longitudinal recording employs ultrathin metallic films (thinner than 100 nm) to prevent the circular magnetization mode. In this case, however, a sawtooth magnetization mode is frequently obtained at the transition, even in very thin and highly coercive films. The effective transition length is given by the sawtooth amplitude and is approximately equal to $L = M_r \delta / H_c$, which usually amounts to one half to one third of the thickness for typical film parameters. It should be noted that the minimum transition length depends on M_r as well as on H_c for all types of longitudinal recording media. This is a distinct disadvantage, because it is difficult to optimize both quantities simultaneously with respect to the transition width. We recall that this problem is absent in perpendicular recording media.

We will conclude this section by briefly discussing the most important magnetic-recording materials currently employed. More details can be found in the surveys of Hibst and Schwab (1994) and Richter (1993). Particulate recording media are most widely used. In these media, magnetic particles are dispersed in an organic binder system. A survey of some important materials used for these magnetic particles is given in Table 13.3.1. The requirement of high bit density on the ultimate tape or rigid disk dictates that the particle size be small. It was mentioned already that, for avoiding the circular mode, it is desirable to have sufficient anisotropy that keeps the magnetization in the film plane of longitudinal recording media.

Not all of the materials listed have a sufficiently high magnetocrystalline anisotropy so that additional shape anisotropy of the particles is required. For this reason, considerable attention is paid in the manufacturing process of the particles to give them an elongated shape. The presence of anisotropy is also needed for the attainment of coercivity. The exact value of the coercivity needed depends on the specific recording system and has to

Table 13.3.1. Property ranges of magnetic recording particles consisting of various materials. After Hibst and Schwab (1994)

Property	γ-Fe$_2$O$_3$	CrO$_2$	CoFe	MP	BaFe
Specific surface area [m^2 g^{-1}]	15–30	15–40	20–50	30–60	25–70
Particle length [nm]	270–500	190–400	150–400	120–300	50–200
Particle volume [$10^{-5} \mu$m^{-3}]	30–200	10–100	5–100	3–45	2–90
Coercivity [kA m^{-1}]	20–35	25–70	35–75	75–160	50–150
Specific saturation magnetization [Am^2 kg^{-1}]	73–75	76–84	70–80	125–170	55–60
Density [10^3 kg m^{-1}]	4.8	4.8	4.8	6.0	5.3

MP: metal particle; BaFe: modified Ba ferrite.

be optimized in the manufacturing process. As a rule, higher recording densities require higher coercivities in order to avoid demagnetizing effects when the written bits are closely spaced. However, the switching field provided by the head during writing is limited so that the coercivities must not be too high. Satisfactory results are generally obtained with coercivities in the range 25–150 kA m^{-1}.

An important property for obtaining a high signal-to-noise ratio is also the remanence of the recording layer. One of the criteria for selecting recording particles is therefore a high specific magnetization and the capability of the particles to be loaded at high volume fractions into the polymeric binder system. Volume fractions close to 40 vol.% should be possible. Higher volume fractions are less desirable because of the high demands in mechanical properties required for the polymer/particle composite medium. Schematic representations of the microstructure in Metal Particle (MP) tapes and Barium Ferrite (BaFe) tapes are displayed in the top part of Fig. 13.3.4.

Magnetic oxides have the advantage of being chemically fairly stable. Their disadvantage is their comparatively low specific magnetization. Much higher specific magnetizations would be obtained when using pure-metal particles. However, the small metal particles are pyrophoric and have to be protected by a passivation layer. The latter is usually obtained during the manufacturing process by means of controlled particle oxidation. This leads to a stable oxide shell when the thickness is about 4 nm, meaning that roughly half of the particle consists of oxide. This is the main reason why the range of specific saturation magnetization values listed in Table 13.3.1 for the MP materials are far below the values of the pure metals. Figure 13.3.4 illustrates that the saturation magnetization of the tape, due to particle passivation and the low volume fraction, has dropped by a factor of about six with respect to the value for pure iron.

Magnetic thin-film media are free of organic binder materials and principally can have much higher remanences than particulate media. Generally, they have thicknesses of only a few hundred nanometers. Even in magnetic thin films prepared by metal evaporation (ME), only a part of the volume is magnetic. This can be seen in the lower part of Fig. 13.3.4. Roughly half of the volume consists of voids, which is a consequence of the vapor-deposition process. However, the amount of oxygen in the film is much lower than in metal-particle films, giving them a substantially higher remanence. A further advantage is the very uniform orientation of the particles, which is hardly achieved with particulate media and which generates favorable switching characteristics.

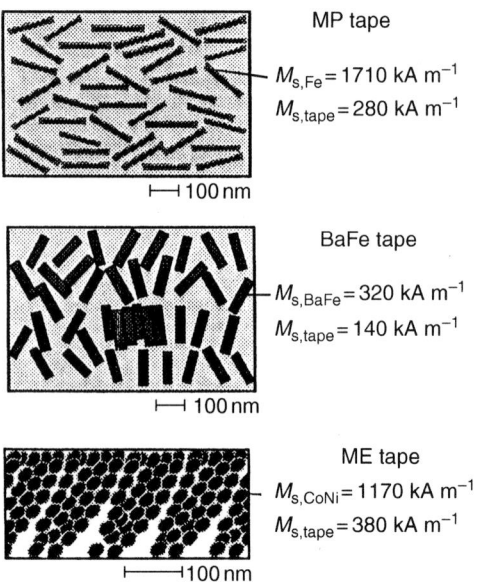

Figure 13.3.4. Schematic representation of the microstructure in tapes based on Metal Particles (MP), Barium Ferrite (BaFe), and a Co–Ni alloy prepared by Metal Evaporation (ME). After Richter (1993).

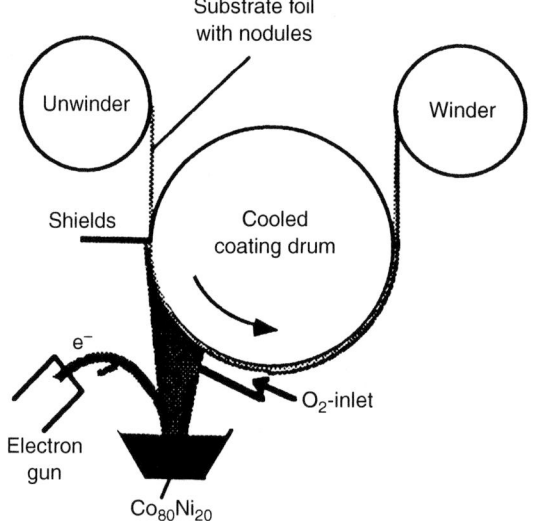

Figure 13.3.5. Schematic diagram showing the production process of thin magnetic films prepared by metal evaporation (ME) of a Co–Ni alloy by means of an electron gun. Controlled oxygen admission serves to magnetically isolate the particles from each other. After Richter (1993).

Thin magnetic films have replaced most of the particulate media in rigid-disk drives and are preferred media in videotape applications. Lodder (1998) has presented a comprehensive review of such media. An illustration of the vapor-deposition process is given in Fig. 13.3.5. The electron gun consists of a hot-metal filament from which electrons are emitted via a high voltage. The beam of electrons is directed into the crucible containing the master alloy via magnetic fields that can deflect this beam. The evaporation rate of the alloy can be adjusted instantaneously by adjusting the power generating the electron beam. The orientation of the crystallites in the film depends on the position of the crucible. In the arrangement shown in the figure, a curved columnar structure of the magnetic layer is obtained. Tapes prepared by this oblique-evaporation technique are used for longitudinal recording. Thin magnetic films used for perpendicular recording are prepared by a symmetrical arrangement of the source relative to the tape. In this case, a columnar structure is obtained with the column axes perpendicular to the film plane. Co–Cr alloys are a preferred medium for perpendicular recording.

References

de Boer, F. R., Boom, R., Mattens, W. C. M., Miedema, A. R., and Niessen, A. K. (1988) in F. R. de Boer and D. G. Pettifor (Eds) *Cohesion in metals*, Amsterdam: North Holland.

Buschow, K. H. J. (1984) in K. A. Gschneidner Jr and L. Eyring (Eds) *Handbook of the physics and chemistry of rare earths*, Amsterdam: North Holland, Vol. 7, p. 265.

Gambino, R. J., Chaudhari, P., and Cuomo J. J. (1973) *AIP Conf. Proc.*, 18, 578.

Hansen, P. (1991) in K. H. J. Buschow (Ed.) *Magnetic materials*, Amsterdam: North Holland, Vol. 6, p. 289.

Hartmann, M. (1982) *PhD Thesis*, Univerity of Osnabrück.

Hibst, H. and Schwab, E. (1994) in R. W. Cahn et al. (Eds) *Materials science and technology*, Weinheim: VCH Verlag, Vol. 3B, p. 211.

Imamura, N. and Mimura, Y. (1976) *J. Phys. Soc. Japan*, 41, 1067.

Lodder, J. C. (1998) in K. H. J. Buschow (Ed.) *Magnetic materials*, Amsterdam: North Holland, Vol. 11, p. 291.

Mee, C. D. and Daniel, E. D. (1987) *Magnetic recording*, New York: McGraw-Hill.

Reim, W. and Schoenes, J. (1990) in K. H. J. Buschow (Ed.) *Magnetic materials*, Amsterdam: North Holland, Vol. 5, p. 133.

Richter, H. J. (1993) in K. H. J. Buschow et al. (Eds) *High density digital recording*, Dordrecht, The Netherlands: Kluwer Academic Publishers, NATO ASI Series E, Vol. 229, p. 197.

Suzuki T. (1984) *IEEE Trans. Magn.*, 20, 675.

14

Soft-Magnetic Materials

14.1. INTRODUCTION

Soft-magnetic materials are mainly used in magnetic cores of transformers, motors, inductors, and generators. Of prime importance for applications in cores are a high permeability, low magnetic losses, and a low coercivity. Definitions of all these quantities are given in Fig. 14.1.1. Other important factors, in particular for large electrical equipment, are a high magnetic flux and low costs.

Unalloyed iron, silicon–iron, and aluminium–iron alloys are widely used in high-power machines. However, for some critical applications, more expensive materials are more suitable. Examples of such materials are Permalloy, Supermalloy, various types of amorphous alloys and nanocrystalline alloys. Several important soft-magnetic materials will be discussed below.

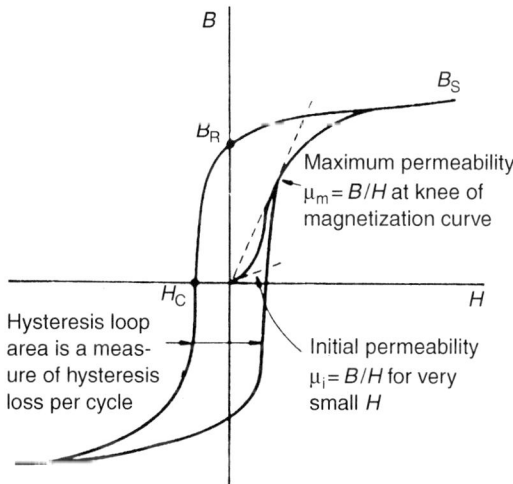

Figure 14.1.1 Hysteresis loop of a magnetic material in which several magnetic quantities relevant to soft-magnetic materials are defined. From Slick (1980).

Domain theory for rotational processes leads to the following expression for the initial permeability:

$$\mu_i = p_\mu \frac{J_s^2}{\mu_0 \langle K \rangle}, \qquad (14.1.1)$$

where J_s is the average saturation magnetization of the material and p_μ is a dimensionless prefactor close to unity. The effective anisotropy constant $\langle K \rangle$ covers all sources of anisotropy energy such as, for instance, the intrinsic magnetocrystalline anisotropy of the material considered and the shape anisotropy.

The coercivity is closely related to the initial permeability because both quantities depend on the effective anisotropy constant $\langle K \rangle$:

$$H_c = p_c \frac{\langle K \rangle}{J_s}, \qquad (14.1.2)$$

where p_c is a dimensionless prefactor close to unity.

A domain-wall-motion model in which the grain size is taken into account leads to the expression:

$$\mu_i \propto \frac{J_s^2 D}{\sqrt{A \langle K \rangle}}, \qquad (14.1.3)$$

where A is the exchange constant and D the grain size. Maximization of the initial permeability requires maximization of J_s and minimization of $\langle K \rangle$. The latter possibility is the one that is exploited most generally.

The minimization of all sources of anisotropy is important when the attainment of high initial permeability is the primary objective. However, a finite but small anisotropy is still desirable for achieving a square or skew hysteresis loop in an assembly of aligned particles. If the magnetization process is performed with the field applied in the easy direction of the aligned anisotropic particles, one obtains a high remanence and the hysteresis loop is square. By contrast, the material exhibits a low remanence and a skew hysteresis loop when it is magnetized perpendicular to the easy direction. Square-loop materials are commonly used in magnetic amplifiers, memory devices, inverters, and converters. Skew-loop materials are primarily used in unipolar pulse transformers.

It will be discussed later that the magnitude and the directional dependence of the various types of anisotropy depend on the composition and the heat treatment. Magnetocrystalline anisotropy has been the most exploited source of anisotropy. Other types are thermomagnetic anisotropy, slip-induced anisotropy, and shape anisotropy. In practice, one aims at the dominance of one particular type of anisotropy by excluding all other sources of anisotropy as far as possible. Of course, this is not necessary if the easy directions originating from two or more types of anisotropy are parallel. A survey of the anisotropy in various Fe-based soft-magnetic materials has been presented by Soinski and Moses (1995).

14.2. SURVEY OF MATERIALS

Iron. Electrical-grade steel is the soft-magnetic material employed in the largest quantities. The annual demand of the electronics industry amounts to several hundred of

SECTION 14.2. SURVEY OF MATERIALS

thousands of tons. The major part of this material is used for the generation and distribution of electrical energy of which the application in motors takes a prominent position.

Fe–Si alloys. Already at the beginning of the 20th century, it was discovered that the addition of a few percent of Si to Fe increases the electrical resistivity and reduces the coercivity. The latter property leads to higher permeability and lower hysteresis losses. The former property is important because it reduces eddy-current losses. The eddy-current losses increase with the frequency squared and can become a major problem in high-frequency applications. The discovery mentioned has led to a widespread application of Fe–Si alloys, although Si addition results in a slight lowering of the saturation magnetization.

The random orientations of the grains in normally cast Fe–Si alloys imply that magnetic saturation can be reached only by applying magnetic fields considerably higher than the coercivity. This limits the useful maximum magnetic flux B to about 1 T. On the other hand, the hysteresis loops of single crystals are nearly rectangular so that only fields slightly higher than the coercivity are required to drive the core to saturation. This fact was used by Goss (1935) in his development of grain-oriented sheets of Fe–Si with considerably improved properties.

Non-grain-oriented sheets or strips are generally hot rolled to a thickness of about 2 mm and then cold rolled to their final thickness. In order to produce sheets with Goss texture, two cold-rolling steps followed by annealing are required after hot rolling. The annealing treatment after the first cold rolling causes recrystallization and sets a defined initial structure for the Goss texture. In the second cold-rolling step, the final thickness is reached. Also this step is followed by annealing leading to recrystallization. After these treatments, high-temperature annealing in a magnetic field leads to oriented grain growth. The ultimate grain-oriented sheets consist of crystallites that have their (110) planes oriented parallel to the plane of the sheet and that have a common [110] direction within this plane. Results of grain-oriented Fe–Si are compared with those obtained on pure Fe in Fig. 14.2.1.

Fe–Ni alloys. Several magnetic alloys, as for instance Ni–Fe alloys, can acquire magnetic anisotropy when annealed below their Curie temperature. Materials having a fairly square hysteresis loop are obtained when the annealing is performed in the presence of an applied magnetic field. The hysteresis loop may become constricted if no field is present. Examples of both types of materials are shown in Fig. 14.2.2.

The anisotropy obtained in a magnetic material by annealing in a magnetic field is called thermomagnetic anisotropy. Its occurrence has been explained by various authors as being due to short-range directional ordering of atom pairs. The magnetic-coupling energy of a pair of atoms generally depends on the nature of the atoms involved (e.g., Fe–Fe, Fe–Ni, Ni–Ni). Detailed studies have shown that it is primarily the concentration of like-atom pairs that is important for the generation of anisotropy in Ni-rich Ni–Fe alloys. Annealing below the Curie temperature in the presence of an applied magnetic field tends to align the coupled pair atoms in a way that they have their moments in the field direction, so as to minimize the free energy. Fast cooling to a sufficiently low temperature then freezes in the directional order obtained. It leads to a uniaxial magnetic anisotropy, the easy axis of the magnetization direction lying in the field direction. Hysteresis loops measured in this same direction are square. By contrast, skew hysteresis loops are obtained when measuring in a direction perpendicular to the direction of the alignment field applied during annealing.

In an unmagnetized piece of a magnetic material, there is no net magnetization because it is composed of an assembly of magnetic domains with different magnetization directions

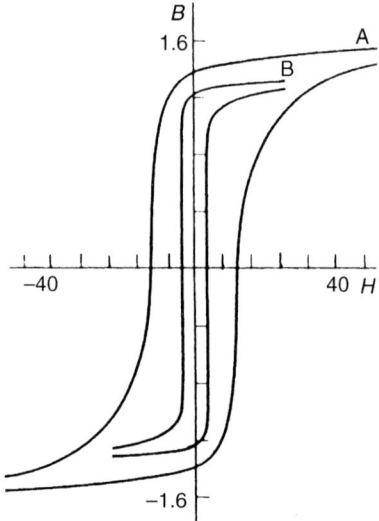

Figure 14.2.1. Hysteresis loop of (A) pure Fe and (B) a grain-oriented alloy consisting of 97% Fe and 3% Si. B and H are given in units of T and A m^{-1}, respectively.

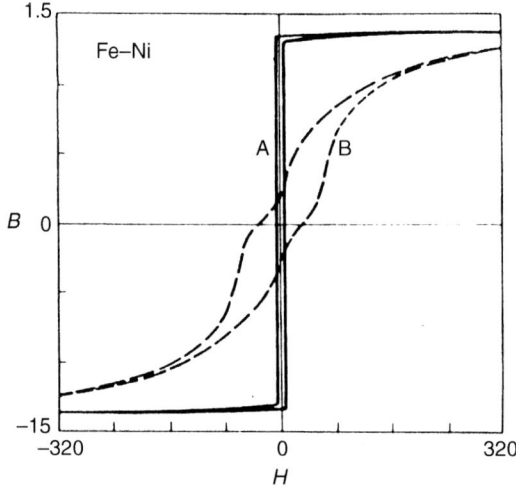

Figure 14.2.2. Hysteresis loop of a Fe–Ni alloy (65% Ni), annealed (A) with and (B) without the presence of an external magnetic field. B and H are given in units of T and A m^{-1}, respectively. From Bozorth (1951).

in a way so as to minimize the magnetostatic energy. An example of such a domain pattern for a single crystal of a cubic material is shown in Fig. 14.2.3.

A domain pattern of a similar nature is also present in a non-magnetized Ni–Fe alloy. When no field is applied during the annealing treatment, the pair moments will become aligned in the local field corresponding to the local magnetization in each magnetic domain.

SECTION 14.2. SURVEY OF MATERIALS

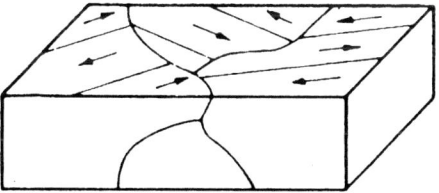

Figure 14.2.3. Domain structure in rolled polycrystalline Ni–Fe alloys. From Chikazumi et al. (1957).

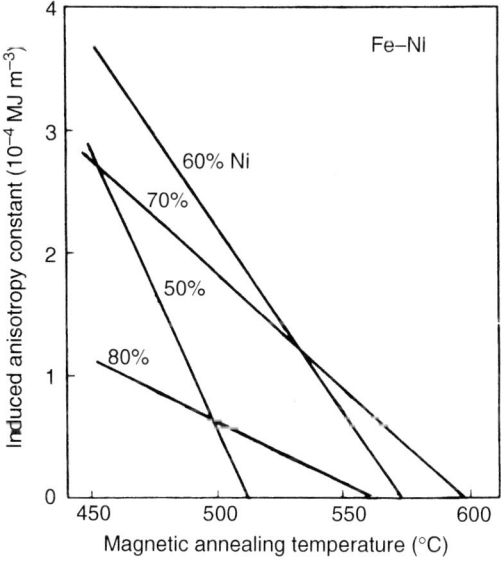

Figure 14.2.4. Dependence on annealing temperature of the thermomagnetic anisotropy constants obtained in Fe–Ni alloys of various concentrations. From Ferguson (1958).

For a polycrystalline alloy, the magnetization directions of the various domains have a random orientation, meaning that the induced anisotropy directions in the various domains will also have a random orientation. This results in a constricted hysteresis loop as shown by means of curve B in Fig. 14.2.2. At this stage, it is good to recall that the main effect of a magnetic field when applied during annealing is to destroy the domain pattern and to align the local magnetization in the field direction across the whole sample.

The value of the thermomagnetic anisotropy constant K_u is generally of the order of a few hundred $J\,m^{-3}$. As can be seen in Fig. 14.2.4, it increases with Fe concentration as a result of an increased number of aligned atom pairs. The value of K_u is generally higher the lower the annealing temperature. More details can be found in the review of Ferguson (1958).

It is important to bear in mind that the thermomagnetic anisotropy is generated by annealing treatments performed below the Curie temperature. Because the pair formation requires diffusion of atoms and because diffusion is a thermally activated process, too low

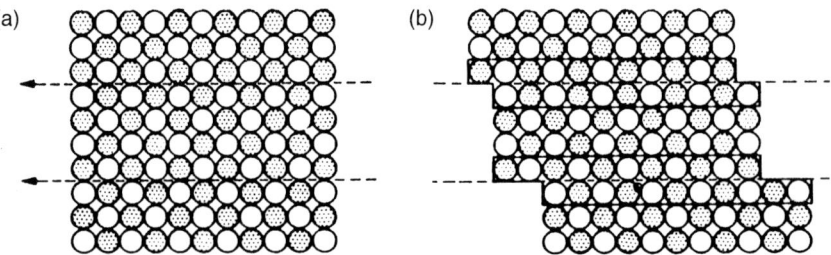

Figure 14.2.5. Schematic representation of slip-induced directional order in a simple alloy. (a) Before slip, (b) after slip on horizontal planes. From Chin and Wernick (1980).

annealing temperatures lead to poor results. That means that large K_u values can only be generated in alloys with sufficiently high Curie temperatures. This is the case for Ni–Fe alloys with Ni content near 65% because at this composition the Curie temperature reaches its maximum in this binary alloy system. Thermomagnetic treatments appear to be less successful in ternary and quaternary alloys in which the Curie temperatures are lower.

Slip- or deformation-induced anisotropy is a second mechanism by means of which the magnetic properties of soft-magnetic materials can be improved (Chin and Wernick, 1980). Also, this type of induced anisotropy depends on directional order of atom pairs, as already discussed above. The difference with thermomagnetically induced anisotropy is that the atomic order is brought about mechanically by means of plastic deformation. Figure 14.2.5 may serve to illustrate the mechanism of slip-induced anisotropy. The atoms are seen to be perfectly ordered before slip (case a), each atom having only dissimilar neighbors. After applying a horizontal sheer stress, the situation has changed (case b). The sheer stress has caused the atoms to slip over one another and has led to the formation of crystallographic defects known as antiphase domain boundaries. Pairs of similar atoms have been created in the vertical direction across the antiphase domain boundaries, whereas the atoms have kept their dissimilar neighbors in the horizontal direction. As in the thermomagnetic case, this directional difference in pair ordering is the origin of the slip-induced anisotropy.

The magnitude of the slip-induced-anisotropy constants are of the order of 10^4 J m^{-3}, which is about 50 times higher than the anisotropy constants obtained by magnetic annealing. The slip-induced-anisotropy constants increase with increasing Fe concentration, as was also found with magnetic annealing. Furthermore, the larger the degree of atomic order prior to deformation, the larger the ultimate anisotropy. This is true in particular for alloys near the Ni$_3$Fe composition. The direction of the easy axis of the slip-induced anisotropy depends on the type of order (long- or short-range), and on the crystal orientation (or texture in the case of polycrystalline material).

Fe–Al and Fe–Al–Si alloys. This is an important group of soft-magnetic materials that are primarily applied in recording heads, to be discussed in the next section. These materials are characterized by high electrical resistivities, high hardness, high permeability, and low magnetic losses. Optimal magnetic properties for the ternary alloys are obtained in a fairly narrow concentration range around 9.6% Si, 5.4% Al, and 85% Fe. This material is also known under the name Sendust.

Soft ferrites. In contradistinction to the hard ferrites discussed in Section 12.7, there exists a group of ferrites that have very low magnetic anisotropy. These materials can be visualized as consisting of mixed oxides and have the general formula $MOFe_2O_3$ with M = Ni, Cu, or Zn. Another group can be described by the formula $[M, Zn]Fe_2O_4$ with M = Cu, Mn, Ni, or Mg. Of particular interest are the ferrites composed of Mn–Zn, Cu–Zn, Cu–Mn, Ni–Zn, Mg–Zn, and Mg–Mn. These materials are primarily used in high-frequency applications where reduction of the various losses accompanying high-frequency magnetization is more important than the static magnetic characteristics. These include head applications to be discussed in a separate section below. A survey of this interesting class of materials has been given by Brabers (1995).

Amorphous alloys. Several types of amorphous alloys have been found to exhibit soft-magnetic properties much superior to those found in crystalline materials. For instance, the core losses measured in amorphous alloys of the composition $Fe_{72}Co_8Si_5B_{15}$ have values that are about an order of magnitude smaller than those of commercial Fe–Si sheets. Most amorphous alloys are prepared by ejecting a molten alloy onto a rotating copper wheel (melt spinning). The high cooling rate associated with this method suppresses crystallization. Amorphous alloys prepared in this manner are also called metallic glasses.

In the amorphous state, the constituting atoms have a more or less random arrangement, grain boundaries being absent. The amorphous state is less stable than the crystalline state and this causes amorphous alloys to spontaneously crystallize upon heating. This amorphous-to-crystalline transformation takes place at the crystallization temperature (T_x) which depends on the composition of the alloy. Most amorphous alloys show a slight atomic rearrangement already at temperatures somewhat below T_x, known as structural relaxation.

As with many crystalline soft-magnetic alloys, after casting or mechanical deformation, a mild thermal treatment is required to remove mechanical stress that can act as a source of coercivity. In amorphous alloys, this stress-release treatment generally does not have the desired result because of the occurrence of structural relaxation. The reason for this is the following. As in crystalline materials, the magnetization processes are governed by nucleation and growth of magnetic domains. This implies that in the remanent state and in the absence of an external magnetic field there will be a distribution of magnetic domains and a corresponding distribution of local magnetization directions. When an amorphous alloy is annealed (below T_x) under these circumstances, structural relaxation will usually be accompanied by an increase in coercivity. This may be illustrated by means of the results shown in Fig. 14.2.6 where the annealed material (curve B) has a substantially higher coercivity than the original melt-spun material (curve A). This increase is a consequence of the presence of magnetic domains with different local magnetization directions, which causes the local structural rearrangements to proceed in a different way.

It is shown in Fig. 14.2.7 how different local-field orientations may lead to different local rearrangements. In the schematic representation in Fig. 14.2.7, it is assumed that pair ordering of the larger type of atoms leads to lower magnetic energy when the axis of the pair of atoms is perpendicular to the local field. The directional ordering in each magnetic domain therefore results in the formation of a local anisotropy. The consequence of this is that the distribution of domains and domain walls during annealing, in the absence of an external field becomes further stabilized and fixed at the original position. These stabilized domain walls cause the mentioned increase in coercivity. In order to be able to stress anneal amorphous alloys under suppression of the undesirable domain-wall fixing

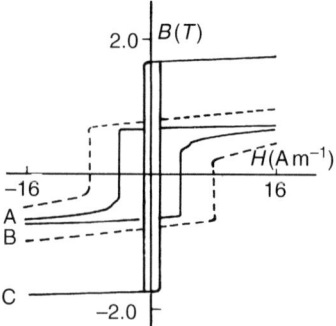

A: As-cast
B: Zero-field annealed
C: Field annealed

Figure 14.2.6. Hysteresis loops observed for an amorphous $Fe_{72}Co_8Si_5B_{15}$ alloy treated in different ways. From Fujimori et al. (1981).

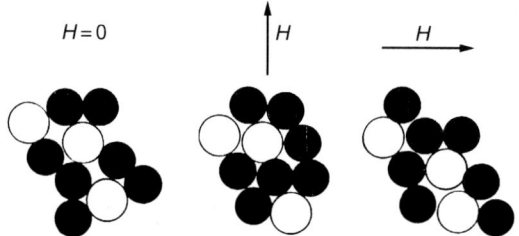

Figure 14.2.7. Atomic rearrangement taking place during structural relaxation of an amorphous alloy in the presence of magnetic fields in two different directions.

due to structural relaxation, one has to destroy the domain pattern by means of an external field during annealing. The beneficial influence of field annealing is shown in Fig. 14.2.6, curve C. Stress release has led to the disappearance of the comparatively large coercivity, whereas the field alignment of the local anisotropies induced by structural relaxation has led to the enhanced remanence. One of the advantages of amorphous alloys is their high electrical resistivity, which leads to low eddy-current losses up to very high frequencies.

It is interesting to compare the effect of magnetic annealing of amorphous alloys with the thermomagnetic treatment of the Fe–Ni alloys discussed above. In both cases, annealing causes changes due to atomic rearrangements. In the Fe–Ni alloys, the corresponding atomic motions proceed by normal diffusion requiring temperatures higher than 450°C. The structural rearrangements in the metastable amorphous alloys occur below 400°C. In both cases, the main effect of the external field is to destroy the domain structure and to align all local fields and hence all thermally induced anisotropies in one direction, that is, in the direction of the external field.

SECTION 14.2. SURVEY OF MATERIALS

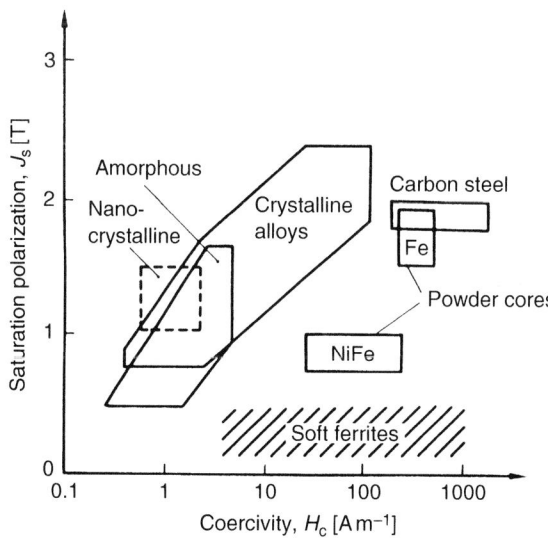

Figure 14.2.8. Comparison of the soft-magnetic properties of several classes of magnetic materials. From Boll (1994).

Nanocrystalline alloys. Nanocrystalline alloys have a microstructure consisting of ultrafine grains in the nanometer range. The first step in the manufacturing of nanocrystalline alloys is the same as used for amorphous alloys. Subsequently, these alloys are given a heat treatment above the corresponding crystallization temperature. The composition of nanocrystalline alloys has been slightly modified with respect to that of soft-magnetic metallic glasses and contains small additions of Cu and Nb. A well-known composition is for instance $Fe_{73.5}Cu_1Nb_3Si_{13.5}B_9$. The effect of the additions is to control the nucleation and growth rates during crystallization. The result is a homogeneous, ultrafine grain structure. In the example mentioned, the grains consist of α-Fe (or rather α-Fe:Si) having a grain diameter of typically 10 nm. This structure leads to relatively high electrical resistivities and makes these alloys suitable for high-frequency applications. In fact, nanocrystalline alloys fill the gap between amorphous alloys and conventional polycrystalline alloys and offer the possibility of tailoring superior soft-magnetic properties for specific applications. In Fig. 14.2.8, the soft-magnetic properties of various groups of materials are compared.

It was mentioned already at the beginning of this chapter that a major requirement for the attainment of superior soft-magnetic properties is generally a low or vanishing magnetocrystalline anisotropy. The magnetocrystalline anisotropy constant K_1 of the ultrafine grains is related to the crystal symmetry; the local easy axis of magnetization being determined by the crystal axis. The anisotropy constant is about $K_1 \approx 10$ kJ m^{-3} for the grains of α-Fe:Si-20at.%, that form the main constituent phase in nanocrystalline $Fe_{73.5}Cu_1Nb_3Si_{13.5}B_9$. This is much too large to explain by itself the low coercivity ($H_c < 1$ A m^{-1}) and the high permeability ($\mu_i \approx 10^5$).

The key to the understanding of the superior soft-magnetic properties of the nanocrystalline alloys mentioned is that the anisotropy contribution of the small, randomly oriented, α-Fe:Si-20 at.% grains is quite substantially reduced by exchange interaction (Herzer, 1989, 1996). The critical scale where the exchange energy starts to balance the anisotropy energy is given by the ferromagnetic-exchange length

$$L_{\text{exch}} = \sqrt{\frac{A}{K_1}}, \qquad (14.2.1)$$

where A represents the average exchange energy as already introduced in Chapter 12. For α-Fe:Si-20 at.%, the value of the exchange length is about $L_{\text{exch}} \approx 35$ nm. The quantity L_{exch} is a measure of the minimum length scale over which the direction of the magnetic moments can vary appreciably. For example, it determines the extent of the domain-wall width, as was discussed in Chapter 12. However, the magnetization will not follow the randomly oriented easy axes of the individual grains if the grain size, D, becomes smaller than the exchange length L_{exch}. Instead, the exchange interaction will force the magnetization of the individual grains to align parallel. The result of this is that the effective anisotropy of the material is an average over several grains and, hence, will strongly reduce in magnitude. In fact, this averaging of the local anisotropies is the main difference with large-grain materials where the magnetization follows the randomly oriented easy axes of the individual grains and where the magnetization process is controlled by the full magnetocrystalline anisotropy of the grains. A more detailed description by means of which one can quantitatively describe this dramatic reduction in anisotropy will be presented for the interested reader in the next section.

14.3. THE RANDOM-ANISOTROPY MODEL

The random-anisotropy model has originally been developed by Alben et al. (1978) to describe the soft-magnetic properties of amorphous ferromagnets. The advent of nanocrystalline magnetic materials has shown, however, that the model is of substantial technical relevance and more generally applicable than considered by Alben. The random-anisotropy model has been applied to nanocrystalline soft-magnetic materials by Herzer (1989, 1996) and the simplified version of the model presented in the review by Herzer (1996) will be followed here.

A schematic diagram representing an assembly of exchange-coupled grains of size D is given in Fig. 14.3.1. The volume fraction of the grains is v_{cr}, and their easy magnetization directions are statistically distributed over all directions. The effective anisotropy constant, $\langle K \rangle$, relevant to the magnetization process of the whole material, can be obtained by averaging the individual grain anisotropies K_1 over the $N = v_{\text{cr}}(L_{\text{exch}}/D)^3$ grains contained within the ferromagnetic-correlation volume $V = L_{\text{exch}}^3$, determined by the exchange length L_{exch}. For a finite number N of grains contained within the exchange volume, there will always be some easiest direction determined by statistical fluctuations. Thus, the averaged anisotropy-energy density is determined by the mean fluctuation amplitude of the anisotropy energy of the N grains, that is,

$$\langle K \rangle \approx \frac{v_{\text{cr}} K_1}{\sqrt{N}} = \sqrt{v_{\text{cr}}} \cdot K_1 \cdot \left(\frac{D}{L_{\text{exch}}}\right)^{3/2}. \qquad (14.3.1)$$

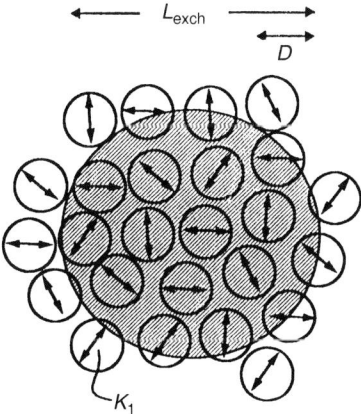

Figure 14.3.1. Schematic diagram representing an assembly of exchange-coupled grains of size D. From Herzer (1996).

As the local magnetocrystalline anisotropies are averaged out this way, the scale on which the exchange interaction dominates expands at the same time. Thus, the exchange length, L_{exch}, has to be renormalized by substituting $\langle K \rangle$ for K_1 in Eq. (14.2.1), that is, L_{exch} is self-consistently related to the averaged anisotropy by

$$L_{\text{exch}} = \sqrt{\frac{A}{\langle K \rangle}}. \tag{14.3.2}$$

After combining Eqs. (14.2.1) and (14.3.2), one finds for grain sizes smaller than the exchange length that the averaged anisotropy is given by

$$\langle K \rangle \approx v_{\text{cr}}^2 K_1 \left(\frac{D}{L_{\text{exch}}}\right)^6 = \frac{v_{\text{cr}}^2 D^6 K_1^4}{A^3}. \tag{14.3.3}$$

It should be borne in mind that this result is essentially based on statistical and scaling arguments. This implies that it is not limited to uniaxial anisotropies, but also applies to cubic or other symmetries.

The most prominent feature of the random-anisotropy model is that it predicts a strong dependence of $\langle K \rangle$ on the grain size. Because it varies with the sixth power of the grain size, one finds for $D \approx L_{\text{exch}}/3$ (grain sizes in the order of 10–15 nm) that the magnetocrystalline anisotropy is reduced by three orders of magnitude (toward a few J m^{-3}). It is this very property, that is, the small grain size and the concomitant strongly lowered anisotropy that gives the nanocrystalline alloys their superior soft-magnetic behavior. Correspondingly, the renormalized exchange length as given by Eq. (14.3.2) reaches values that fall into the μm-regime. This is almost two orders of magnitude larger than the natural exchange length as given by Eq. (14.2.1). This has as a further consequence that the domain-wall width, discussed in Section 12.3, can become fairly large in these nanocrystalline materials. It has already been mentioned briefly in Section 13.2 that magnetic domains of different magnetization direction can be optically distinguished from each other by using plane-polarized light and a polarization microscope. High-resolution Kerr-effect studies made on nanocrystalline $Fe_{73.5}Cu_1Nb_3Si_{13.5}B_9$ have confirmed the presence of very wide domain walls of about 2 μm in thickness.

If there are no other forms of anisotropies present, both the coercivity and the initial permeability depend on the randomized effective anisotropy constant $\langle K \rangle$ and are closely related via Eqs. (14.1.1) and (14.1.2). It is important to realize that these relations normally apply to magnetization processes governed by coherent magnetization rotation. According to an argument given by Herzer (1996), these relations are also applicable to magnetization processes proceeding by domain-wall displacements for cases in which $D \ll L_{\mathrm{exch}}$. In fact, on the scale of the nanocrystalline grains (10 nm), the magnetization vector appears to rotate coherently if a domain wall with a width of 2 μm ($\approx L_{\mathrm{exch}}$) passes by.

14.4. DEPENDENCE OF SOFT-MAGNETIC PROPERTIES ON GRAIN SIZE

The grain-size dependence of the magnetic properties of various types of soft-magnetic materials is compared in Fig. 14.4.1. The random-anisotropy model apparently provides a good description of the magnetic properties for grain sizes below about $L_{\mathrm{exch}} \approx 40$–50 nm. The D^6 dependence derived in the preceding section is well reflected in the coercivity and the initial permeability. This implies that Rayleigh's constant, which is proportional to μ_i/H_c, varies as $1/D^{12}$. If the grain size becomes equal to the exchange length, the magnetization process is determined by nearly the full magnetocrystalline anisotropy K_1.

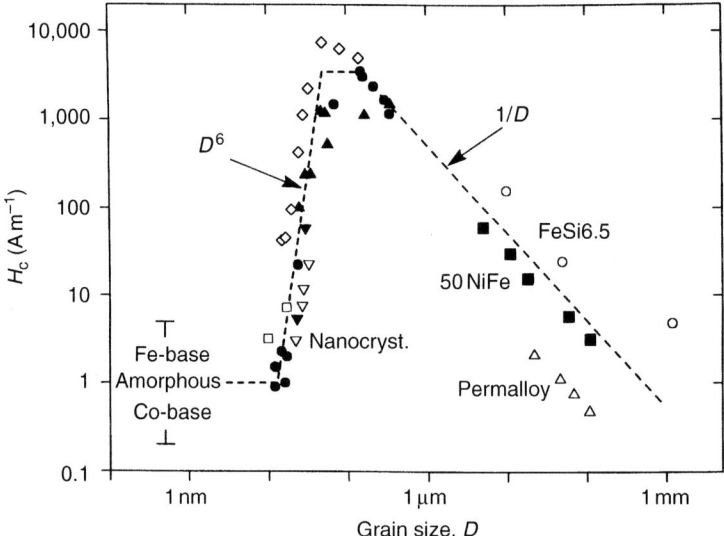

Figure 14.4.1. Comparison of the grain-size dependence of the magnetic properties of various types of soft-magnetic materials. From Herzer (1996).

Accordingly, H_c and $1/\mu_i$ are seen to pass through a maximum in this grain-size regime. When the grain size has eventually become so large that it exceeds the domain-wall width, domains can be formed within the grains. As a consequence, H_c and $1/\mu_i$ tend to decrease again according to the well known $1/D$ law (see Eq. 14.1.3).

14.5. HEAD MATERIALS AND THEIR APPLICATIONS

14.5.1. High-Density Magnetic-Induction Heads

A conventional inductive recording head consists of a slit toroid of a high-permeability material wound by several conductor turns. A schematic representation is shown in Fig. 14.5.1.1. The output voltage V of the head is determined by Faraday's law (Eq. 8.7) and hence by the flux changes due to the medium when passing along the slit. However, in the setup shown in the figure, also the field $H(x, y, z)$ produced by a current i passing through the head windings is of influence. It can be shown that the following expression holds for the output voltage V (Mee and Daniel, 1990):

$$V = -\frac{\mu_0}{i} \int_{\text{vol}} \vec{H}(x, y, z) \cdot \frac{\partial \vec{M}(x, y, z)}{\partial t} dV$$

$$= \frac{\mu_0 v}{i} \int_{\text{vol}} \vec{H}(x, y, z) \cdot \frac{\partial \vec{M}(x, y, z)}{\partial x} dV, \quad (14.5.1.1)$$

where $M(x, y, z)$ is the magnetization of the medium, and v is the medium velocity in the x direction.

It follows from Eq. (14.5.1.1) that the output voltage depends on the velocity v of the medium relative to the head. This implies that the larger the speed of the medium, higher is the sensitivity. In some applications where a high sensitivity and a high storage density are required (video applications and several audio and data-processing applications) one

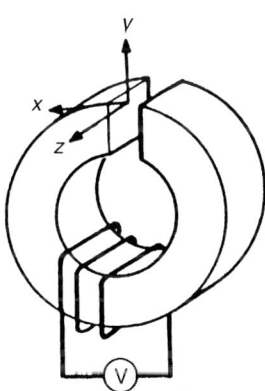

Figure 14.5.1.1. Schematic representation of a conventional inductive-recording head. From Mee and Daniel (1990).

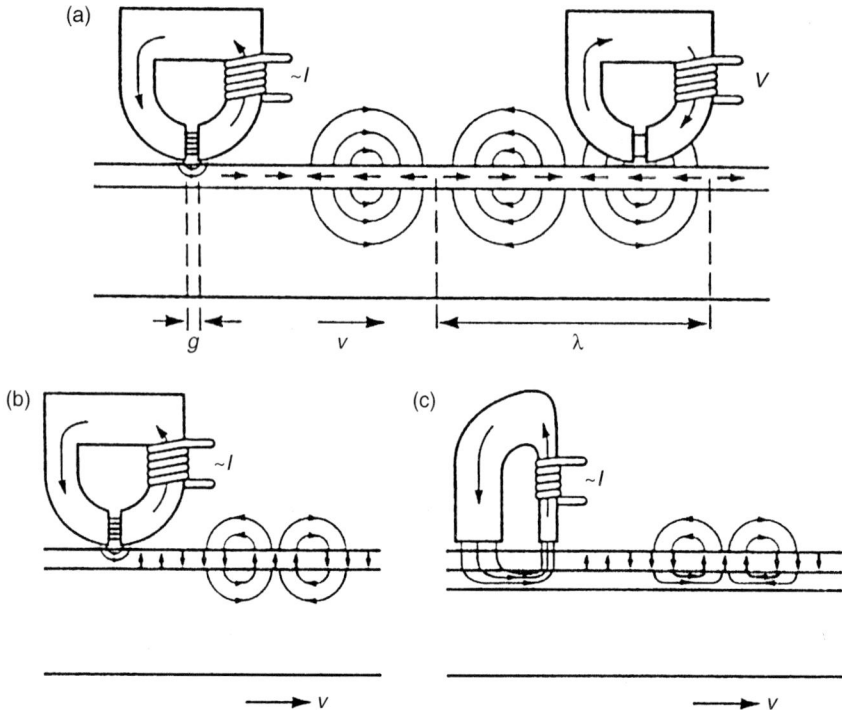

Figure 14.5.1.2. Side view of the recording process. (a) Longitudinal recording with a ring head, (b) perpendicular recording with a ring head and a single-layer medium, (c) perpendicular recording with a single-pole head and a double-layer medium. From Bernards and Schrauwen (1990).

therefore does not employ stationary heads but rotating heads. Heads in modern magnetic storage systems are designed in a way that they can develop a hydrodynamic and self-acting air bearing under steady operating condition, which minimizes the head–medium contact. There is only physical contact between the medium and the head during the starts and stops.

In modern data-storage tape and disk drives, the head-to-medium separation is of the order of 0.1–0.3 mm, the head and medium surfaces have roughnesses of the order of 2–10 nm. The need for higher recording densities requires that the surfaces be as smooth as possible and the flying heights as low as possible. A schematic representation of a recording process is shown in Fig. 14.5.1.2.

In general, one may distinguish between two types of heads, magnetic inductive heads and magnetoresistive heads. There are two different physical principles involved in these heads. Consequently also the material requirements for the two types are different. In the next two sections, both types of materials will be briefly discussed.

Soft-magnetic materials are widely employed for the fabrication of magnetic recording heads. These materials must have a high saturation magnetization in order to produce a large gap field. A high permeability is required in order to ensure high efficiency and a small magnetostriction in order to ensure low medium-contact noise. The coercivity has to be low in order to ensure a low thermal noise, and a high electrical resistivity in order to reduce

Table 14.5.1.1. Properties of several materials used for magnetic inductive heads (from Fedeli, 1993)

Characteristic property	Ni–Zn ferrite	Mn–Zn ferrite	Mn–Zn ferrite	NiFe alloy	FeAlSi alloy	Amorphous CoNbZrTa
Curie temp., °C	150–200	90–300	100–265	580	500	505
Saturation induction, T	0.4–0.45	0.4–0.6	0.4	0.8	0.9–1.1	0.8
Coercive field, Oe	0.1–0.4	0.1–0.2	0.05	0.2	—	—
Permeability $\times 10^{-2}$ at 1 MHz	8	7–10	4–10	10	13–20	>50
10 MHz	3	4	—	4	4	20
100 MHz	—	—	—	—	—	—
Magnetostriction	$\cong 0$	$\cong 0$	$\neq 0$	$\cong 0$	$\simeq 0$	$\neq 0$
Resistivity, $\mu\Omega$ cm	$\cong 10^{11}$	$\cong 5 \times 10^6$	$>5 \times 10^5$	20–60	85–150	120
Thermal expansion coeff. $\times 10^6$, deg^{-1}	9	11	11	13	11–15	11.5
Vickers hardness	850	800	—	120	500–700	900
Wear resistance	exc.	exc.	exc.	poor	good	exc.
Thermal stability	exc.	exc.	exc.	good	good	fair

eddy currents. To ensure good reliability and a long operating life, the materials must exhibit a good thermal stability and a high resistance to wear and corrosion. Table 14.5.1.1 lists a number of materials used for inductive-head applications.

14.5.2. Magnetoresistive Heads

In the early 1970s, a novel type of reading heads was introduced, based on several types of transition-metal alloys such as Ni–Fe, Ni–Co, and Co–Fe. The working principle of these heads is the magnetoresistive effect, which entails a decrease of the electrical resistivity when the direction of the applied current is rotated away from the magnetization direction.

The magnetoresistive heads have characteristics that are fundamentally different from those of the inductive heads described in the preceding section. In its simplest form, the head consists of a narrow sensor strip of height h and width w, mounted in a plane perpendicular to the moving recording medium. It is connected to leads at each end carrying a sense current I as shown in Fig. 14.5.2.1.

Due to the magnetoresistive effect, the electrical resistivity of each portion of this strip depends on the angle θ between the direction of magnetization $\vec{M}$ and the current-density vector $\vec{j}$:

$$\rho = \rho_0 + \Delta\rho \cos^2 \theta. \quad (14.5.2.1)$$

In most of the conventional transition-metal alloys, the values of $\Delta\rho/\rho_0$ are 2–6%. Values of about an order of magnitude higher can be reached in special alloys consisting of small ferromagnetic single-domain particles in a non-magnetic metallic medium (granular films) High values of $\Delta\rho/\rho_0$ are also reached in multilayer films. Multilayer thin films and granular

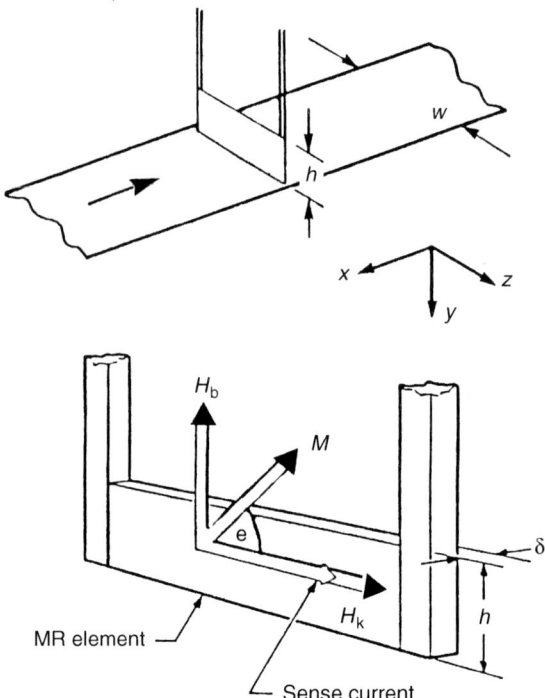

Figure 14.5.2.1. Schematic representation of a magnetoresistive sensor strip of width w and height h placed perpendicular to the moving medium tape (top). From Fedeli (1993). Enlarged view of the magnetoresistive sensor strip (bottom).

thin films are currently indicated as materials giving rise to giant magnetoresistance (GMR) effects.

In most applications of vertical magnetoresistive heads, one has $\theta = 0$ in the quiescent state, owing to the magnetic anisotropy. In Fig. 14.5.2.1, this anisotropy is represented by the anisotropy field H_k. During operation of the head, the magnetization vector $\vec{M}$ will rotate over an angle θ given by

$$\sin\theta = \frac{H_y}{H_k + H_d}, \qquad (14.5.2.2)$$

where H_y is the sum of the field ($\pm h_y$) emanating from the recording medium and a bias field H_b applied to linearize the response of the head to the field of the medium. The field H_d in Eq. (14.5.2.2) accounts for local demagnetization effects of the magnetoresistive element, the latter effects being more pronounced at the edges of the magnetoresistive element. Assuming that the bias and demagnetizing fields are constant, and that only the field from the medium h_y depends on y and z, one finds for the output voltage

$$V = \text{constant} \times lw\Delta\rho. \qquad (14.5.2.3)$$

The signal levels of magnetoresistive heads are much higher than that of conventional inductive heads. Furthermore, the output signal of the magnetoresistive head depends only on the instantaneous fields of the media, and hence is independent of the media velocity or the time rate of change of the fields. This offers a significant advantage for reading low-velocity media. Here, we recall that the sensitivity of inductive reading is strongly dependent on the relative velocity between head and medium, since this type of recording is based on Faraday's law (see Section 14.5.1).

References

Alben, R. J., Becker, J. J., and Chi, M. C. (1978) *J. Appl. Phys.*, 49, 1653.
Bernards, J. P. C. and Schrauwen, C. P. G. (1990) *Thesis*, University of Twente, The Netherlands.
Boll, R. (1994) in R. W. Cahn et al. (Eds) *Materials science and technology*, Weinheim: VCH Verlag, Vol. 3B, p. 401.
Bozorth, R. M. (1951) *Ferromagnetism*, New York: Van Nostrand.
Brabers, V. A. M. (1995) in K. H. J. Buschow (Ed.) *Magnetic materials*, Amsterdam: North Holland Publ. Co., Vol. 8, p. 189.
Chikazumi, S., Suzuki, K., and Iwata, H. (1957) *J. Phys. Soc. Japan*, 12, 1259.
Chin, G. Y. and Wernick, J. H. (1980) in E. P. Wohlfarth (Ed.) *Ferromagnetic materials*, Amsterdam: North Holland Publ. Co., Vol. 2, p. 55.
Fedeli, J. M. (1993) in K. H. J. Buschow, G. J. Long and F. Grandjean (Eds) *High density digital recording*, Dordrecht: Kluwer Academic Publ., NATO ASI Series E, Vol. 229, p. 251.
Ferguson, E. T. (1958) *J. Appl. Phys.*, 29, 252.
Fujimori, H., Yoshimoto, H., and Matsumoto, T. (1981) *J. Appl. Phys.*, 52, 1893.
Goss, N. P. (1935) *Trans. Am. Soc. Met.*, 23, 511.
Herzer, G. (1989) *IEEE Trans. Magn.*, 25, 3327.
Herzer, G. (1996) in K. H. J. Buschow (Ed.) *Magnetic materials*, Amsterdam: North Holland Publ. Co., Vol. 10, p. 415.
Hibst, H. and Schwab, E. (1994) in R. W. Cahn et al. (Eds) *Materials science and technology*, Weinheim: VCH Verlag, Vol. 3B, p. 211.
Mee, C. D. and Daniel, E. D. (1990) *Magnetic recording handbook*, New York: McGraw-Hill.
Slick, P. I. (1980) in E. P. Wohlfarth (Ed.) *Ferromagnetic materials*, Amsterdam: North Holland Publ. Co., Vol. 2, p. 189.
Soinski, M. and Moses, A. J. (1995) in K. H. J. Buschow (Ed.) *Magnetic materials*, Amsterdam: North Holland Publ. Co., Vol. 8, p. 189.

15

Invar Alloys

The origin of thermal expansion is the presence of anharmonic terms in the potential energy expression describing the mutual separation of a pair of atoms at a temperature T. If x represents the displacement of the atoms from their equilibrium position, the potential energy may be written as

$$V(x) = cx^2 - gx^3 - fx^4. \tag{15.1}$$

The term in x^3 is a measure of the asymmetry of the mutual repulsion of the atoms, and the term in x^4 can be regarded as describing the general softening of the vibrations at large amplitudes.

In order to calculate the average displacement, we will follow Kittel (1953) and use the Boltzmann distribution function (analogous to Eqs. 3.1.3 and 3.1.4), which weights the possible values of x with a factor representing their thermodynamic probability.

$$\bar{x} = \frac{\int_{-\infty}^{\infty} xe^{-V(x)/kT} dx}{\int_{-\infty}^{\infty} e^{-V(x)/kT} dx}. \tag{15.2}$$

For small displacements, the anharmonic contribution to the potential energy is relatively small. In this case, the integrands may be expanded as

$$\int xe^{-V(x)/kT} dx \cong \int e^{-cx^2/kT} \left[x + \frac{gx^4}{kT} + \frac{fx^5}{kT} \right] dx = \frac{3\sqrt{\pi}}{4} \left(\frac{g}{kT} \right) \left(\frac{kT}{c} \right)^{5/2}$$

and

$$\int e^{-V(x)/kT} dx \cong \int e^{-cx^2/kT} dx = \left(\frac{\pi kT}{c} \right)^{1/2},$$

so that

$$\bar{x} = \frac{3gkT}{4c^2}. \tag{15.3}$$

This result shows that the temperature coefficient of the thermal expansion is a constant. In classical mechanics, the mean value of the energy E of an oscillator in the harmonic approximation is equal to kT. For this reason, one may also write Eq. (15.3) in the form

$$\bar{x} = \frac{3g\bar{E}}{4c^2}, \tag{15.4}$$

which suggests that the approximate quantum-mechanical result would be obtained by substituting for $\bar{E}$ the corresponding quantum-mechanical expression for the energy of a harmonic oscillator with frequency ω. This then leads to

$$\bar{x} = \frac{3g}{4c^2} \cdot \frac{\hbar\omega}{e^{\hbar\omega/kT} - 1}. \tag{15.5}$$

It is useful to bear in mind that the specific heat as well as the thermal-expansion coefficient are temperature derivatives of E, which means that the thermal-expansion coefficient is proportional to the specific heat.

On the basis of Eq. (15.5), one would furthermore expect the thermal-expansion coefficient to decrease rather abruptly when the temperature falls below the characteristic temperature of the oscillator and to go to zero if the temperature goes to zero kelvin. This is what is commonly observed. The third law of thermodynamics requires that the thermal-expansion coefficient vanishes if the temperature goes to zero.

A schematic representation of the thermal-expansion behavior expected, if only the lattice anharmonicity contributes, is shown in Fig. 15.1 (dashed-dotted curve). In many magnetic materials, the thermal expansion takes quite a different form, as shown for instance by the full curve in the same figure. The total thermal expansion can be subdivided into a lattice contribution (ω_l) and a contribution due to magnetic effects. The latter contribution is called the spontaneous volume magnetostriction (ω_s) and is indicated by the broken curve in Fig. 15.1.

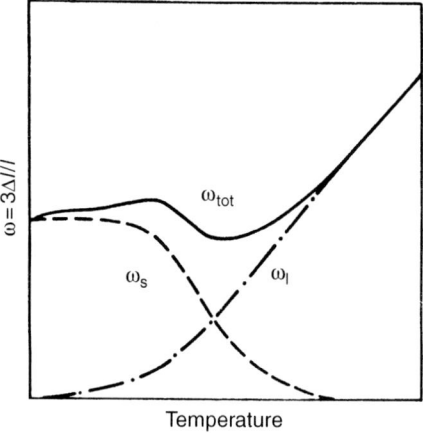

Figure 15.1. Schematic representation of the thermal expansion ω_{tot} of an Invar-type of material. The total effect can be separated into the lattice contribution (ω_l) and the spontaneous volume magnetostriction (ω_s).

CHAPTER 15. INVAR ALLOYS

The most conventional treatment of magnetovolume effects is based on a model in which the magnetic moments are assumed to be localized. The magnetic contribution to the volume change can be represented by the two-spin correlation function $\langle \vec{m}_i \cdot \vec{m}_j \rangle$ via

$$\omega_s = \kappa \sum_{i,j} C_{\text{loc}} \langle \vec{m}_i \cdot \vec{m}_j \rangle, \quad (15.6)$$

where the summation is taken over all magnetic sites, and where κ is the compressibility. The quantity C_{loc} is the magnetovolume coupling constant. It originates from the volume dependence of the exchange constant J_{ij}, responsible for the magnetic coupling between two magnetic moments i and j (see, for instance, Eq. 4.4.2). This means that C_{loc} is proportional to dJ_{ij}/dV.

In the case of alloys or intermetallic compounds based on 3d metals, one has to realize that the 3d electrons occupy a narrow energy band having a width of a few electron volts, as has been discussed in Section 7.1. In order to describe magnetovolume effects in these materials, it is therefore necessary to take the band character of these electrons into consideration. The reason for this is that there is an intimate connection between interatomic distances, bandwidth and magnetic properties, as will be further discussed below.

It was outlined in Section 7.2 that the spin polarization of the 3d band that causes the formation of magnetic moments is a trade-off between exchange energy (which is gained) and the kinetic energy (which is lost). However, the increase in kinetic energy required for the band polarization can be kept low if this band polarization is accompanied by volume expansion. This may be seen as follows. If volume expansion occurs, one expects a concomitant decrease of the bandwidth W on the basis of Eq. (7.1.1). It can be easily verified by means of Fig. 7.1.1 that the expenditure in kinetic energy required to realize a given 3d-band polarization (i.e., to realize a given amount of electron transfer from the minority band to the minority band) will be lower, the smaller the bandwidth (i.e., the higher the density of states).

To a first approximation, the increase in kinetic energy is proportional to the square of the magnetization. The volume change due to band polarization can therefore be written as

$$\omega_{\text{band}} = \kappa C_{\text{band}} M^2, \quad (15.7)$$

where C_{band} represents the magnetovolume coupling constant associated with the band character of the 3d electrons.

At low temperatures ($T \ll T_C$) the spin-correlation function $\langle m_i \cdot m_j \rangle$ in Eq. (15.6) may be approximated by M_s^2, so that the total volume magnetostriction can now be written as

$$\omega_{\text{band}} = \kappa(C_{\text{band}} + C_{\text{loc}})M_s^2. \quad (15.8)$$

Well-known materials with Invar properties are alloys of iron and nickel in a concentration range close to the composition Fe_3Ni. It is interesting and instructive to compare the Invar properties of these alloys with results of calculations of their electronic band structure. The volume dependence of the total energies of non-magnetic and ferromagnetic states derived from these calculations (Williams et al., 1983) is shown in Fig. 15.2. In fcc FeNi (top part), the ferromagnetic state is the ground state, having an energy lower than the paramagnetic state. The situation for fcc Fe is shown in the bottom part of the

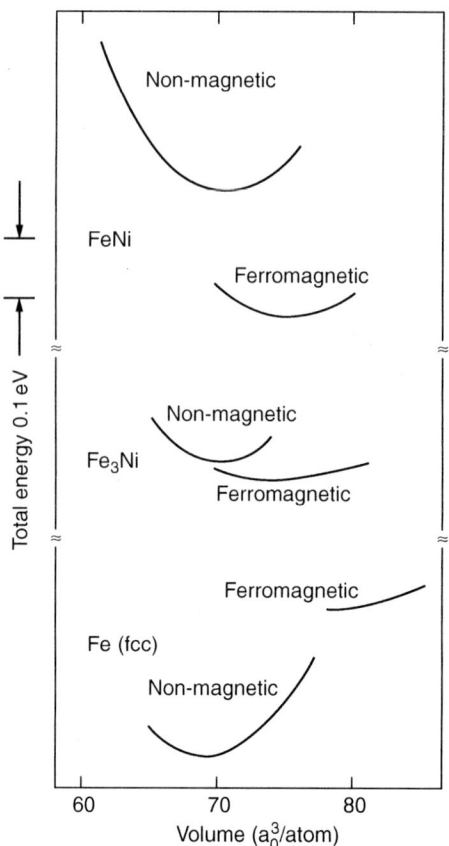

Figure 15.2. Volume dependence of the total energy calculated for pure Fe and ordered Fe–Ni compounds, spanning the Invar-concentration range. The total-energy difference between the ferromagnetic and non-magnetic state is seen to change sign in this concentration range. Non-magnetic means Pauli paramagnetic, or not spin-polarized, or the absence of local magnetization. The calculations are based on the local-spin-density treatment of exchange and correlation. The augmented-spherical-wave method was employed for the required spin-polarized self-consistent energy-band calculations. After Williams et al. (1983).

figure. It is seen that here the non-magnetic state has the lower energy. Computational results for the Invar alloy Fe_3Ni are shown in the middle part of the figure. There is not much difference in energy between the non-magnetic state and the ferromagnetic state. At low temperatures, only the ferromagnetic state will be populated, having its minimum energy at a comparatively high volume. Williams and co-workers ascribe the Invar properties to thermal excitations into the non-magnetic state for which the energy minimum is seen to occur at a significantly lower volume. Increasing temperature, therefore, leads to a gradual loss of the spontaneous volume expansion associated with the ferromagnetic state.

Invar alloys are employed in many devices for which a low thermal expansion is desirable. A detailed description of the physics and application of Invar alloys is presented in the

CHAPTER 15. INVAR ALLOYS

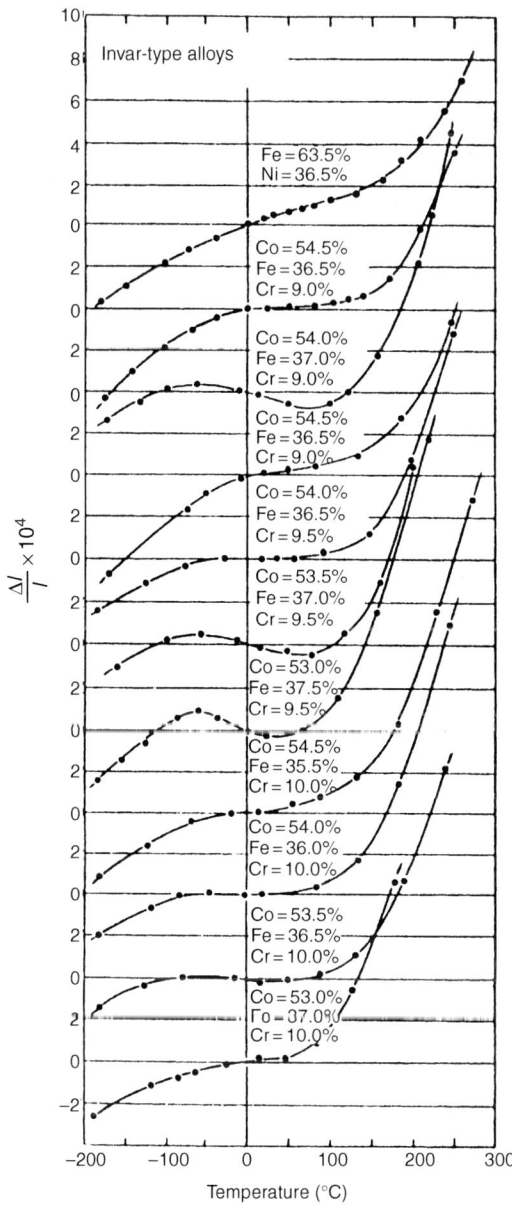

Figure 15.3. Thermal-expansion behavior of several stainless-steel Invar alloys. After Kaya (1978).

surveys of Kaya (1978), Wasserman (1991), and Shiga (1994). The properties of a number of Invar alloys based on stainless steel are shown in Fig. 15.3. Invar properties are also found in many intermetallic compounds. For example, compounds of the $R_2Fe_{14}B$ type, discussed extensively in Section 12.5, also display such properties (Fig. 15.4).

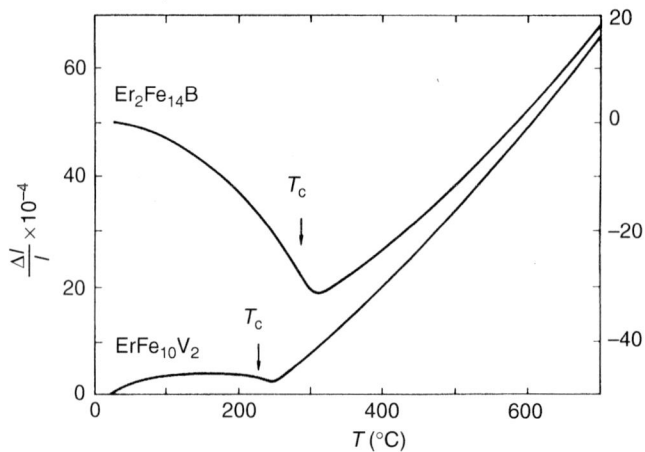

Figure 15.4. Thermal-expansion behavior of two intermetallic Fe-based rare-earth compounds.

References

Kaya, S. (1978) *Physics and application of Invar alloys*, Tokyo: Maruzen Co.
Kittel, C. (1953) *Introduction to solid state physics*, New York: John Wiley.
Shiga, M. (1994) Invar alloys, in R. W. Cahn et al. (Eds) *Materials science and technology*, Weinheim: VCH Verlag, *Vol. 3B*, p. 159.
Wasserman, E. F. (1991) Moment-volume instability in transition metal alloys, in K. H. J. Buschow (Ed.) *Ferromagnetic materials*, Amsterdam: North Holland, *Vol. 5*, p. 237.
Williams, A. R., Moruzzi, V. L., Gelatt Jr., C. D., and Kübler, J. (1983) *J. Magn. Magn. Mater., 10*, 120.

16

Magnetostrictive Materials

Magnetostriction can be defined as the change in dimension of a piece of magnetic material induced by a change in its magnetic state. Generally, a magnetostrictive material changes its dimension when subjected to a change of the applied magnetic field. Alternatively, it undergoes a change in its magnetic state under the influence of an externally applied mechanical stress. By far the most common type of magnetostriction is the Joule magnetostriction where the dimensional change is associated with a distribution of distorted magnetic domains present in the magnetically ordered material. It is well known that ferromagnetic and ferrimagnetic materials adopt a magnetic domain structure with zero net magnetization in the demagnetized state in order to reduce the magnetostatic energy. In a material showing Joule magnetostriction, each of the magnetic domains is distorted by interatomic forces in a way so as to minimize the total energy.

Concentrating on a single of these domains, for materials with positive (negative) magnetostriction, the dimension along the magnetization direction is increased (decreased) while simultaneously the dimension in the direction perpendicular to the magnetization direction is decreased (increased), keeping the volume constant. This means that for a piece of magnetostrictive material, consisting of an assembly of many magnetostrictively distorted domains, one expects dimensional changes when an external field causes a rotation of the magnetization direction within a domain, and/or when the external field causes a growth of domains, for which the magnetization direction is close to the field direction, at the cost of domains for which the magnetization direction differs more from the field direction. We will return to this point later.

The magnetostrictive properties will reflect the symmetry of the crystal lattice when the piece of material is a single crystal. In this case, the length changes observed at magnetic saturation depend on the measurement direction as well as on the initial and final direction of the magnetization of the single crystal. As shown in more detail in several reviews (Cullen et al., 1994; Gignoux, 1992; Andreev, 1995), frequently only two magnetostrictive constants are required to describe the fractional length change associated with the saturation magnetostriction (λ_s) in cubic materials:

$$\frac{\Delta l}{l} = \frac{3}{2}\lambda_{100}\left(\alpha_x^2\beta_x^2 + \alpha_y^2\beta_y^2 + \alpha_z^2\beta_z^2 - \frac{1}{3}\right) + 3\lambda_{111}(\alpha_x\alpha_y\beta_x\beta_y + \alpha_y\alpha_z\beta_y\beta_z + \alpha_x\alpha_z\beta_x\beta_z). \tag{16.1}$$

In this expression, α and β represent the direction cosines with respect to the x, y, and z crystal axes of the magnetization direction and the length-measurement direction, respectively. This relation makes it possible to describe the magnetostrictive properties for any choice of the latter two directions if the two magnetostrictive constants λ_{100} and λ_{111} are available. These two magnetostriction constants have the following physical meaning: λ_{100} (λ_{111}) represents the change in length or saturation magnetostriction in the $\langle 100 \rangle$ ($\langle 111 \rangle$) direction when the magnetization direction is also along the $\langle 100 \rangle$ ($\langle 111 \rangle$) direction after the material has been cooled through its Curie temperature.

In the following, we will consider the macroscopic properties of a cubic ferromagnetic material for which the preferred magnetization direction is along $\langle 100 \rangle$. When a large single crystal of this material is cooled to below the Curie temperature, it will be in the unmagnetized state by adopting a magnetic-domain structure that reduces its magnetostatic energy. The magnetization in each of these domains is along one of the $\langle 100 \rangle$ directions and, if $\lambda_{100} > 0$, each of these domains is elongated in the corresponding $\langle 100 \rangle$ direction. However, no distortion will be observed upon cooling to below the Curie temperature because the distribution of $\langle 100 \rangle$ directions in the domain structure leads to a cancelation of the distortion. This may be illustrated by means of Fig. 16.1. In this figure, we have assumed for simplicity that only domains are present in which the preferred direction is along cubic directions of the type [100] or [010]. The situation changes drastically if we apply a magnetic field along one of these cubic directions, say [100]. The single crystal now has become one single domain with the magnetization along the field direction. No cancellation of distortive contributions is possible and the single crystal has become elongated along the field direction. In other words, when applying a magnetic field along one of the main crystallographic directions of a magnetically ordered but unmagnetized piece of cubic material, we can produce an elongation or shrinking. Which of these latter two possibilities is realized depends on the sign of the magnetostriction constant in this particular direction.

In tetragonal or hexagonal materials, one frequently encounters easy-axis anisotropy, the preferred magnetization direction being along the crystallographic c direction. In that case, the domain structure will consist of domains separated by 180° walls. Because of

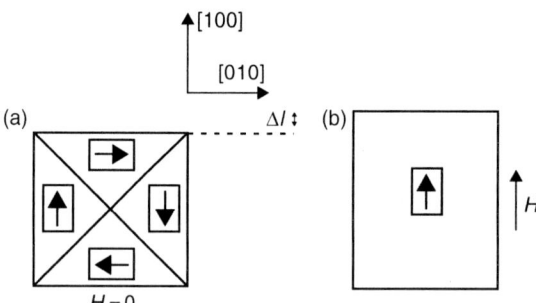

Figure 16.1. (a) Schematic representation of a possible domain structure in a single crystal of a cubic material in which the easy magnetization directions are along the $\langle 100 \rangle$ directions. The tetragonal distortions due to magnetostriction ($\lambda_{100} > 0$) of the various domains have been indicated by rectangles. (b) Field-induced elongation Δl of the crystal due to domain-wall motion, leading to the disappearance of the domains of which the magnetization direction is different from the applied-field direction. After Gignoux (1992).

Table 16.1. Magnetostriction constants of several cubic materials at room temperature

Material	λ_{100} (10^{-6})	λ_{111} (10^{-6})
Fe	24	−22
Ni	−51	−23
TbFe$_2$	—	2460
SmFe$_2$	—	−2100

the equivalence of the positive and negative c direction, domains on either side of the domain wall will experience the same type of deformation in the magnetically ordered state. This means that no special effect will be observed when applying a magnetic field in one of these directions, causing the disappearance of domains that have their magnetization in the opposite direction. Therefore, cubic materials are generally considered to be more appropriate for obtaining magnetostriction effects generated by domain-wall motion. The magnetostriction constant of several cubic materials can be compared with each other in Table 16.1.

In polycrystalline materials, the situation is more complex than in single crystals because one has to relate the magnetostriction of the whole piece of material to the magnetoelastic and elastic properties of the individual grains. This problem cannot be solved by an averaging procedure. For this reason, it is assumed that the material is composed of a large number of domains with the strain uniform in all directions. It can be shown that, for a material in which there is no preferred grain orientation, this leads to the expression (Chikazumi, 1966):

$$\lambda_s = \tfrac{2}{5}\lambda_{100} + \tfrac{3}{5}\lambda_{111}. \tag{16.2}$$

Inspection of the data listed in Table 16.1 shows that in particular the cubic compound TbFe$_2$ (also called Terfenol) has quite outstanding magnetostrictive properties. For this reason, this compound has found applications in magneto-mechanical transducers. It can, for instance be used to generate field-induced acoustic waves at low frequencies in the kHz range (Sonar). Alternatively, its changes in magnetic properties under external stresses have led to applications in sensors for force or torque. A variety of other magnetostrictive materials and their properties are discussed in the reviews of Cullen et al. (1994) and Andreev (1995).

The microscopic origin of magnetostrictive effects has sometimes been attributed to dependencies of the exchange energy or the magnetic dipolar energy on interatomic spacing. However, these approaches proved less satisfactory because they were not able to account for the magnitude of the observed magnetostriction. As discussed in more detail by Morrish (1965), it is more likely that magnetostriction has the same origin as the magnetocrystalline anisotropy. In that case, magnetostriction can be viewed as arising because the spontaneous straining of the lattice lowers the magnetocrystalline energy more than it raises the elastic energy. Indeed, the analysis of modern magnetostrictive materials based on rare earths (R) and 3d metals (T) has shown that there is an intimate connection between magnetostriction and crystal-field-induced anisotropy, as is explained in more detail in the treatments of Clark (1980), Morin and Schmitt (1990), and Cullen et al. (1994). Generally, the theoretical framework describing magnetostrictive effects is fairly complex. We will restrict ourselves therefore to a simplified discussion of these effects as given by Gignoux (1992).

Inspection of the crystal-field Hamiltonian presented in Eq. (5.2.7) shows that strain effects can be introduced via strain dependence of the crystal-field parameters A_n^m that characterize the surrounding of the aspherical 4f-electron charge cloud. The lowest order magnetoelastic effects depend on the derivative of these parameters with respect to strain, which leads to supplementary terms in the Hamiltonian that couple strains with the second-order Stevens operators. It gives rise to isotropic as well as to anisotropic distortions of which the latter have magnetic symmetry and are dominant. For instance, Morin and Schmitt (1990) have shown that the magnetoelastic-energy term associated with the tetragonal-strain mode (γ), and hence with λ_{100}, reads as:

$$H_{me}^\gamma = -B^\gamma(\varepsilon_1^\gamma O_2^0 + \sqrt{3}\varepsilon_2^\gamma O_2^2), \tag{16.3}$$

where B^γ is a magnetoelastic coefficient and the ε^γ are strain components of the corresponding symmetry. When calculating the magnetoelastic energy at finite temperatures, one has to form thermal averages $\langle O_n^m \rangle$ of the Stevens operators. These thermal averages are generally small above the magnetic-ordering temperature in rare-earth–transition-metal compounds, but can adopt appreciable values below T_C. Figure 16.2 presents a very simple example illustrating the physical principles behind magnetoelastic effects. Here, a simple ferromagnetic rare-earth compound has been chosen where normally the 4f-charge cloud does not have an electric quadrupolar moment in the paramagnetic state. In this case, the cubic crystal field leads to energy levels whose 4f orbitals correspond to a cubic distribution of the 4f electrons, as displayed in the left part of the figure. The magnetic symmetry is tetragonal below T_C when one of the fourfold axes is the easy magnetization direction.

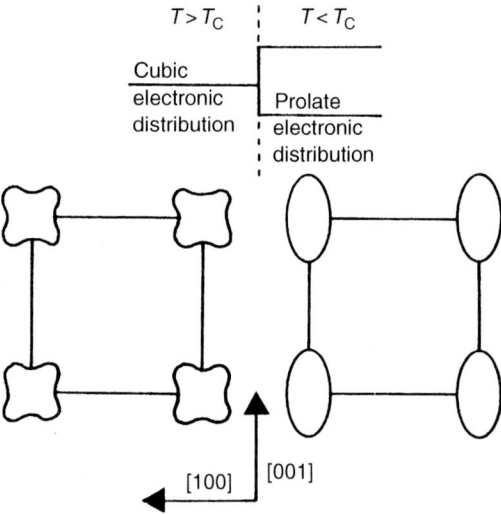

Figure 16.2. Diagram illustrating the occurrence of spontaneous tetragonal distortion due to a change from a cubic electron distribution to a prolate electron distribution when a cubic ferromagnetic material is cooled to below its Curie temperature. After Gignoux (1992).

The second-order crystal-field term introduced by this symmetry leads to a ground state with a 4f-electron distribution that is no longer cubic. If one assumes, for instance, a prolate shape, the coupling to the strain mode gives rise to a lattice expansion along the [001] direction and a contraction along [100] and [010]. For a different sign of the magnetostriction constant λ_{100}, one would have observed a lattice contraction along [001] and an expansion along [100] and [010].

References

Andreev, A. V. (1995) in K. H. J. Buschow (Ed.) *Magnetic materials*, Amsterdam: Elsevier Science Publ., *Vol. 8*, p. 59.

Chikazumi, S. (1966) *Physics of magnetism*, New York: John Wiley and Sons.

Clark, A. E. (1980) in E. P. Wohlfarth (Ed.) *Ferromagnetic materials*, Amsterdam: North Holland, *Vol. 1*, p. 531.

Cullen, R., Clark, A. E., and Hathaway, Kristl. B. (1994) in R. W. Cahn et al. (Eds) *Material science and technology*, Weinheim: VCH Verlag, *Vol. 3B*, p. 529.

Gignoux, D. (1992) in R. W. Cahn et al. (Eds) *Material science and technology*, Weinheim: VCH Verlag, *Vol. 3A*, p. 367.

Morin, F. and Schmitt, D. (1990) in K. H. J. Buschow (Ed.) *Magnetic materials*, Amsterdam: Elsevier Science Publ., *Vol. 5*, p. 1.

Morrish, A. H. (1965) *The physical principles of magnetism*, New York: John Wiley and Sons.

Author Index

Alben, R. J., 156
Andreev, A. V., 171, 173

Barbara, B., 25, 45
Becker, R., 25
Beckman, O., 22
Bethe, H., 20, 21, 22
Boer, F. R. de, 134
Boll, R., 155
Bozorth, R. M., 150
Brabers, V. A. M., 153
Brooks, M. S. S., 41, 71, 72, 73
Buschow, K. H. J., 24, 39, 108, 121, 136

Charap, S. H., 25, 29
Chikazumi, S., 25, 29, 102, 151, 173
Chin, G. Y., 152
Clark, A. E., 173
Clarke, J., 89
Clegg, A. G., 102, 105, 123
Coehoorn, R., 53
Cohen, E. R., 83
Cullen, R., 171, 173

Daniel, E. D., 142, 159
Danielsen, O., 116
Durst, K. D., 100
Duzer, T. van, 89

Fedeli, J. M., 161, 162
Ferguson, E. T., 151
Franse, J. J. M., 70, 102
Friedel, J., 66, 68, 69
Fujimori, H., 154

Gambino, R. J., 133
Gaunt, P., 99
Giacomo, P., 83
Gignoux, D., 32, 33, 171, 172, 174

Givord, D., 115
Goldfarb, R. B., 83
Gorter, E. W., 38
Goss, N. P., 149
Guillot, M., 122

Hansen, P., 136
Hartmann, M., 133, 136, 137
Heine, V., 63
Henry, W. E., 14
Herring, C., 21
Herzer, G., 156, 157, 158
Hilscher, G., 83
Hibst, H., 143
Hofmann, J. A., 93
Hutchings, M. T., 45, 46

Imamura, N., 138

Johansson, B., 41, 71, 72, 73

Kaya, S., 169
Kittel, C., 44, 165
Kools, F., 123
Koon, N. C., 41
Kronmüller, H., 100, 113, 115

Lindgard, P. A., 116
Little, W. A., 95
Liu, J. P., 41
Lodder, J. C., 144
Lundgren, L., 22

Marcon, G., 125
Martin, D. H., 17, 25, 59
McCaig, M., 102, 109, 123
Mee, C. D., 142, 159
Mimura, Y., 138

Morin, F., 173, 174
Morrish, A. H., 17, 25, 29, 92, 173
Moses, A. J., 148

Nicklow, R. M., 41

Pauling, L., 69

Radwanski, R. J., 70
Ram, V. S., 99
Reim, W., 136
Rhyne, J. J., 41
Richter, H. J., 142, 144
Rudowicz, C., 116

Schmitt, D., 173, 174
Schoenes, J., 136
Schwab, E., 142, 143
Shiga, M., 169
Slater, J. C., 20, 21, 22, 69
Slick, P. I., 144
Smit, J., 98

Soinski, M., 148
Sommerfeld, A., 20
Stapele, R. P. van, 39
Stoner, E. C., 65
Strnat, K. J., 99
Sucksmith, W., 99
Suzuki, T., 140, 141

Thompson, J. E., 99
Tokuhara, K., 119
Turner, C. W., 89

Verhoef, R., 38
Vos, K. J. de, 126, 128

Wasserman, E. F., 169
Wernick, J. H., 152
White, R. M., 25, 45
Wijn, H. P. J., 98
Williams, A. R., 167, 168

Zijlstra, H., 85, 87, 115

Subject Index

adiabatic demagnetization, 94
alloys of rare earths and 3d metals, 70
Alnico, 107
Alnico alloys, 124
Alnico DG, 127
amorphous alloys, 131, 133, 153
amorphous Gd–Fe films, 133
anisotropic distortions, 172
anisotropic magnet, 126
anisotropic sintered magnet, 121
anisotropy constants, 97, 115
anisotropy energy, 97
anisotropy field, 97, 99
anisotropy of magnetization directions, 54
annealing in a magnetic field, 150
antibonding states, 71
antiferromagnet, 26
antiferromagnetic interactions, 22
antiferromagnetism, 26
aspherical 4f-electron charge cloud, 56
asymptotic Curie temperature, 23
axial quadrupole moment, 56

Ba ferrite, 113, 143
Barium Ferrite (BaFe) tapes, 144
$BaFe_{12}O_{19}$, 122
B–H curve, 106
Bethe–Slater curve, 20
bias sputtering, 134
Bloch walls, 110
Bohr magneton, 4
Boltzmann distribution, 11
bonding states, 71
Brillouin function, 11, 28

caloric effects in magnetic materials, 91
charge cloud of 4f electrons, 56
circular magnetization mode, 140
Co–Cr films, 141, 145
coercive fields, 82
coercivity, 82, 106

coercivity mechanisms, 112
columnar-crystallized Alnico, 128
compensation temperature, 38, 135
compensation point, 38
coupling between the 3d and 4f spins, 72, 73
CrO_2, 143
crystal field, 43
crystalline electric field, 43
crystal-field Hamiltonian, 45
crystal-field interaction, 44
crystal-field-induced anisotropy, 54, 115
crystal-field-split states, 47
crystal-field parameters, 50
crystal-field potential, 45
crystal-field-split eigenstates, 50
crystal-field-split levels, 52
Curie constant, 15
Curie law, 13
Curie temperature, 19
Curie–Weiss law, 25

deformation-induced anisotropy, 152
De Gennes factor, 8
demagnetization factor, 79
demagnetizing fields, 78
density of electron states, 63
density of states, 68
diamagnetism, 59
die-pressing of magnetic materials, 121
domain walls, 109
domains, 109
domain-wall-motion model, 148

easy-magnetization direction, 97
effective exchange energy, 63
effective moments, 15
electrical-grade steel, 149
electric-field gradient, 56
electronic specific heat, 92
entropy, 93
erasability of written information, 132
exchange-coupled grains, 156

exchange length, 156
exchange interaction, 20
excited multiplet levels, 15
expectation value of the 4f radius, 46

Faraday effect, 136, 146
Faraday method, 86
Fe-based intermetallic compounds, 21
Fe–Al and Fe–Al–Si alloys, 152
$Fe_{73.5}Cu_1Nb_3Si_{13.5}B_9$, 158
Fe_2O_3, 143
Fermi level, 63
Fe–Si alloys, 149
ferrimagnetic compounds, 39
ferrimagnetism, 34
ferromagnetic-exchange length, 156
ferromagnetic materials, 19
ferromagnetism, 22
Ferroxdure, 117
field lines, 78
flux density, 76
flux lines, 78

g-factor for the free electron, 5
giant magnetoresistance, 162
Goss texture, 149
Gouy method, 85
grain-oriented Fe–Si, 149
ground-state multiplet level, 6

hard ferrites, 122
head materials, 159
head-to-medium separation, 160
Heisenberg exchange interaction, 19
hexaferrites, 122
HFFP method, 40
high-density recording, 131
high-field susceptibility, 69
high-density magnetic recording materials, 139
Hund's rules, 7
hysteresis loops, 81, 105

inhomogeneous nucleation, 114
initial permeability, 147
intermetallic compounds, 21, 70
intersublattice coupling, 70
intersublattice interaction, 37
intersublattice-coupling constant, 40
intersublattice-molecular-field constant, 26
intrasublattice interaction, 37
intrasublattice-molecular-field constant, 26
intrinsic coercivity, 80
Invar alloys, 165
Invar properties, 168
ionic properties of iron-group elements, 9
ionic properties of the rare-earth elements, 8
irreversible losses, 109

isotropic magnets, 126
itinerant-electron magnetism, 63
itinerant electron systems, 63

Kerr effect, 132, 136
Kerr rotation, 136

Langevin formula for diamagnetism, 60
Langevin function, 25
Landé spectroscopic g-factor, 7
lanthanide series, 8
Lenz's law, 59
ligand ion, 44
liquid-phase sintering, 121
longitudinal magnetic recording, 140
longitudinal recording, 139
Lorentz force, 59
low-field susceptibility, 113

magnetically ordered state, 19
magnetic anisotropy, 97
magnetic cores, 147
magnetic entropy, 94
magnetic-induction heads, 159
magnetic losses, 109, 147
magnetic-ordering, 21
magnetic permeability in vacuum, 75
magnetic polarization, 80
magnetic properties of iron-group elements, 9
magnetic properties of the rare-earth elements, 8
magnetic quantum number, 3
magnetic recording, 139
magnetic-recording medium, 139
magnetic splitting of the ground-multiplet, 16
magnetic susceptibility, 15, 65
magnetic tape, 143
magnetic thin-film media, 144
magnetization reversal, 112
magnet materials, 106
magnetocaloric effect, 91
magnetoelastic effects, 174
magnetoelastic-energy, 174
magnetometer, 87, 89
magneto-optical recording, 131
magneto-optical recording materials, 133
magneto-optical rotation, 136
magnetoresistive effect, 161
magnetoresistive heads, 161
magnetoresistive recording-head, 161
magnetostriction, 171
magnetostrictive constants, 171, 173
magnetostrictive materials, 173
magnetostrictive properties, 173
magnetovolume coupling constant, 167
majority band, 68
manufacturing route for permanent
    magnets, 119
magnetovolume effects, 165

mass susceptibility, 78
materials for high-density magnetic recording, 143
maximum energy product, 105
Maxwell's equations, 79
measurement techniques, 85
metallic thin films, 144
metal particle (MP) tapes, 144
metamagnetic transition, 34
M-type ferrites, 122
minority band, 67
miscibility gap, 125
molar susceptibility, 78
molecular field, 20
multiplet, 6
$[M, Zn]Fe_2O_4$ with $M = $ Cu, Mn, Ni or Mg, 153

nanocrystalline alloys, 155, 158
nanocrystalline soft-magnetic materials, 155, 158
$Nd_2Fe_{14}B$, 119
$Nd_2Fe_{14}B$ permanent magnets, 119
Néel temperature, 27
Ni–Fe alloys, 149
nucleation field, 113
nucleation of Bloch walls, 113
nucleation-type magnet, 114

oblique-evaporation technique, 144
operator equivalents, 46
optical recording, 131
orbital-angular-momentum quantum number, 3
orbital states of electrons, 3

pair ordering, 152
pair-ordering model, 134
paramagnetic Curie temperature, 23, 27
paramagnetism of free ions, 11
particulate media, 140, 143
Pauli's principle, 4
permalloy, 147
permanent magnets, 105
permanent-magnet materials, 117
perpendicular magnetic recording, 140
pinning-controlled coercivity, 114
pinning-type magnets, 114
point-charge approximation, 52
point-charge model, 45
preferred magnetization directions, 54, 97
preferred moment direction, 57
principal quantum number, 3
production route for $Nd_2Fe_{14}B$ permanent magnets, 119
propagation field for Bloch walls, 113

radius of 4f electron charge cloud, 46
random-anisotropy model, 156
rare-earth-based magnet materials, 119

rare-earth-based magnets, 119
rare-earth series, 15
RCo_5, 118
read-out of written bits, 132, 135, 136
recoil energy, 108
recoil line, 108
recoil product, 108
recording head, 159, 160
recording process, 160
reduced magnetization, 25
reduced matrix elements, 46
reduced temperature, 25
$R_2Fe_{14}B$ compounds, 117
remanence, 105
rigid disk, 142
rigid-disk drives, 145
Russell–Saunders coupling, 6

saturation magnetostriction, 171
second-order crystal-field parameter, 116
second-order Stevens factor, 57
self-consistent energy-band calculations, 168
shape anisotropy, 127
shape of the 4f-charge cloud, 56
short-range ordering, 93
single-domain particles, 159
sintered magnet, 119
sintered magnet bodies, 121
SI units, 79
skew hysteresis loop, 149
Slater–Pauling curve, 69
slip-induced anisotropy, 152
$SmCo_5$, 117, 118
$Sm(Co,Fe,Cu,Zr)_7$, 118
soft ferrites, 153
soft-magnetic materials, 147
specific heat, 91
specific-heat anomaly, 91
specific-heat discontinuity at Curie temperature, 92
spectroscopic splitting factor, 5
spin-down band, 63
spin flop, 33
spin-correlation function, 167
spinodal decomposition, 124
spin-orbit interaction, 6
spin polarization of the 3d band, 64
spin quantum number, 4
spin-reorientation temperature, 118
spin states of electrons, 3
spin-up band, 63
spontaneous magnetization, 19
spontaneous straining of the lattice, 173
spontaneous volume magnetostriction, 166
sputtered Cd–Co films, 134
SQUID magnetometer, 89
statistical average of magnetic moments, 12
Stevens' operator equivalents, 46
Stoner criterion for ferromagnetism, 65
Stoner enhancement factor, 66
Stoner–Wohlfarth model, 127

strong ferromagnetism, 68
Sucksmith–Thompson plots, 99
superexchange interaction, 122
supermalloy, 147
susceptibility balance, 85
susceptibility enhancement, 65

thermal expansion, 165
thermal-expansion coefficient, 166
thermomagnetic anisotropy, 151
thermomagnetic writing of bits, 131
Ticonal XX, 108
total orbital angular momentum, 5
total spin angular momentum, 6
torque, 30
torque magnetometer, 101

ultrathin metallic films, 142
unit of magnetic field strength, 75

unit of magnetization, 77
unit of the magnetic induction, 76
units, 75

valence-electron asphericities, 53
vapour deposition of thin magnetic films, 134, 144
vector model of atoms, 4, 5
vertical recording, 140
vibrating-sample magnetometer, 87
volume magnetostriction, 166
volume susceptibility, 78

wall energy, 111
wall pinning, 114
wall thickness, 111
weak ferromagnetism, 68
Weiss field, 19
Weiss constant, 23
working point of a magnet, 105